New York State Coach, Gold Edition, Science, Grade 4

Triumph Learning®

New York State Coach, Gold Edition, Science, Grade 4
272NY
ISBN-10: 1-60471-721-1
ISBN-13: 978-1-60471-721-1

Contributing Writer: Kathy Fox
Cover Image: Falling apple. The apple is the state fruit of New York. © Jupiter Images.

Triumph Learning® 136 Madison Avenue, 7th Floor, New York, NY 10016

© 2010 Triumph Learning, LLC

All rights reserved. No part of this publication may be reproduced in whole or in part, stored in a retrieval system, or transmitted in any form or by any means, electronic, mechanical, photocopying, recording or otherwise, without written permission from the publisher.

Printed in the United States of America.

Table of Contents

			Major Understandings
Letter to the Student		6	
Test-Taking Checklist		7	
New York State Learning Standards Correlation Chart		8	
New York State Core Curriculum Science Skills		22	

Unit 1 Analysis, Inquiry, and Design

Chapter 1	**Scientific Inquiry**	23	
Lesson 1	Observe and Ask Questions	24	S1.1a, b; S1.2a; IS3
Lesson 2	Identify Systems and Patterns	29	M2.1b; S2.3a; PS1.1c; CT1, 4, 5
Lesson 3	Tools for Observing and Measuring	33	M3.1a; S2.3a; PS3.1e; CT3, 4, 5
Lesson 4	Measuring Systems	37	M1.1a; S2.3a; PS3.1d
Lesson 5	Safety in Science	41	S2.3a
	Chapter 1 Review	44	
Chapter 2	**Experiments**	47	
Lesson 6	Design an Investigation	48	S1.3a; S2.1a; S2.3a; IP2
Lesson 7	Collect and Organize Data	53	M3.1a; S2.3a, b; S3.1a; IS 1; CT2, 5
Lesson 8	Figure Out Data	57	M2.1a; S2.3a; S3.2a; S3.3a; IS 3; CT6
Lesson 9	Present Information to Others	62	S2.3a; S3.3a; S3.4a, b; IS 1, 2
	Chapter 2 Review	66	
Chapter 3	**Engineering Design**	69	
Lesson 10	Technology and Engineering	70	T1.1a, b; IS3
Lesson 11	Study a Design	74	T1.1a, b; T1.2a, b, c; IP1
Lesson 12	Figure Out New Ideas	78	T1.1c; T1.3a, b, c; T1.4a, b; IS 2; CT2, IP1, 2
Lesson 13	Test Your Ideas	82	T1.5a, b, c; CT2, 6; IP 1
	Chapter 3 Review	85	

Duplicating any part of this book is prohibited by law.

				Major Understandings
Unit 2		**The Physical Setting**		
Chapter 4		**Earth's Movement in Space**	89	
	Lesson 14	Day and Night	90	PS1.1a, c
	Lesson 15	Seasons	94	PS1.1a
	Lesson 16	Our View of the Moon	98	PS1.1a
	Lesson 17	Measuring Time	102	PS1.1b
		Chapter 4 Review	106	
Chapter 5		**Planet Earth**	109	
	Lesson 18	Weather	110	S2.3a; PS2.1a, b; PS3.1e
	Lesson 19	The Water Cycle	115	PS2.1c; LE6.2c
	Lesson 20	How Earth's Land Changes	119	PS2.1d
	Lesson 21	Extreme Natural Events	123	PS2.1e
		Chapter 5 Review	128	
Chapter 6		**Matter**	133	
	Lesson 22	Physical Properties of Matter	134	PS3.1a, b, c, e; PS3.2a; CT4
	Lesson 23	Solids, Liquids, and Gases	138	PS2.1c; PS3.2a, b, c
	Lesson 24	Measuring Properties	143	M1.1a, b, c; M3.1a; PS3.1c, e, g
	Lesson 25	Classifying Objects	148	S2.3a; PS3.1f
		Chapter 6 Review	152	
Chapter 7		**Energy**	155	
	Lesson 26	Forms of Energy	156	PS4.1a, b; PS4.1g
	Lesson 27	Heat and Temperature	160	M3.1a; PS4.1a, b, c, f, g
	Lesson 28	Sound and Light	164	PS3.1g; PS4.1a, b, d, g
	Lesson 29	Electrical Circuits	169	PS4.1a, b, c, d, e, g
	Lesson 30	Energy Changing Form	174	M2.1b; PS4.1d; PS4.2a, b
		Chapter 7 Review	177	
Chapter 8		**Forces and Motion**	181	
	Lesson 31	Motion	182	M1.1a; M2.1a; S2.3a; PS5.1a, b
	Lesson 32	How Forces Affect Motion	186	M2.1b; PS5.1b, c, d; PS5.2a
	Lesson 33	Magnets	190	PS5.1e; PS5.2a, b
	Lesson 34	Simple Machines	194	PS5.1d, f
		Chapter 8 Review	199	

Unit 3		The Living Environment	
Chapter 9		**Characteristics of Living Things** 203	
	Lesson 35	What it Means to Be Alive 204	LE1.1a, b, c, d; LE1.2a; LE4.2a, b; LE5.1a; LE5.2a
	Lesson 36	Inherited, Acquired, and Learned Traits 208	LE2.1a, b; LE2.2a, b; LE5.2f; LE6.1e
	Lesson 37	Plant Structures and Functions 212	LE3.1b, c; LE5.1b; LE5.2a; CT1
	Lesson 38	Life Cycles of Plants 217	LE4.1a, b, c, d
	Lesson 39	Animal Structures and Functions 221	LE3.1a, c; LE5.1b; LE5.2b, c, d, e; CT1
	Lesson 40	Animal Life Cycles. 225	LE4.1a, e, f, g
	Lesson 41	Human Health 229	LE5.3a, b
		Chapter 9 Review 234	
Chapter 10		**Living Things and Their Environments** 239	
	Lesson 42	Producers and Consumers 240	S2.3a; LE6.1a, b, c, d; LE6.2a, b
	Lesson 43	Changes in Populations 244	M2.1b; LE3.2a, b; LE5.2g; LE6.1b, e, f
	Lesson 44	Human Changes in the Environment 248	LE7.1a, b, c
		Chapter 10 Review 253	

Investigation 1: How Does Water Change? 257

Investigation 2: Why Does the Moon's Shape Seem to Change? 265

Glossary 273

Pretest 283

Posttest 299

Letter to the Student

Dear Student,

Welcome to *Coach*! This book provides instruction and practice that will help you master the skills and information you need to know. The book also gives you practice answering the kinds of questions you will see on your state's test.

The *Coach* book is organized into units, chapters, and lessons, and includes a Pretest and Posttest. Before you begin the first chapter, your teacher may want you to take the Pretest, to identify areas in which you need improvement. Then you can study the lessons that focus on those areas. Or you can begin with the first chapter of the *Coach* and work through to the end.

Each lesson explains a topic that is important in your study of science. Diagrams and charts will help you understand the information. You can check your understanding by answering the questions at the end of the lesson. The Chapter Reviews contain questions like those you may find on your state test. Answering these questions will help you know which lessons you need to review.

The *Coach* also includes two science investigations. By doing the investigations, you will be practicing skills that all scientists use. The thinking skills you use in science will help you in your study of other subjects, too.

After you have finished the lessons and reviews, you can take the Posttest to see how much you have improved. Even if you did well on the Pretest, you will probably do better on the Posttest because practice makes perfect.

We wish you lots of success this year, and hope the *Coach* will be a part of it!

Test-Taking Checklist

Here are some tips to keep in mind when taking a test. Take a deep breath. You'll be fine!

✓ Read the directions carefully. Make sure you understand what they are asking.

✓ Do you understand the question? If not, skip it and come back to it later.

✓ Reword tricky questions. How else can the question be asked?

✓ Try to answer the question before you read the answer choices. Then pick the answer that is the most like yours.

✓ Look for words that are **bolded**, *italicized*, CAPITALIZED, or underlined. They are important.

✓ Always look for the main idea when you read. This will help you answer the questions.

✓ Pay attention to pictures, charts, and graphs. Do you understand the information in them? Sometimes they can give you hints.

✓ If you are allowed, use scrap paper. Take notes and make sketches if you need to.

✓ Always read all the answer choices first. Then go back and pick the best answer for the question.

✓ Be careful marking your answers. Make sure your marks are clear.

✓ Double-check your answer sheet. Did you fill in the right bubbles?

✓ Read over your answers to check for mistakes. Only change your answer if you're sure it's wrong. Your first answer is usually right.

✓ Work at your own pace. Don't go too fast, but don't go too slow either. You don't want to run out of time.

Good Luck!

New York State Learning Standards Correlation Chart

Learning Standards	Key Ideas	Performance Indicators	Major Understandings	Coach Lesson(s)
STANDARD 1: Analysis, Inquiry, and Design Students will use mathematical analysis, scientific inquiry, and engineering design, as appropriate, to pose questions, seek answers, and develop solutions.					
STANDARD 1 Analysis, Inquiry, and Design **MATHEMATICAL ANALYSIS (M)**	*Key Idea 1:* Abstraction and symbolic representation are used to communicate mathematically.	M1.1 Use special mathematical notation and symbolism to communicate in mathematics and to compare quantities, express relationships, and relate mathematics to their immediate environment.	M1.1a Use plus, minus, greater than, less than, equal to, multiplication, and division signs	4, 24, 31
			M1.1b Select the appropriate operation to solve mathematical problems	24
			M1.1c Apply mathematical skills to describe the natural world	24
	Key Idea 2: Deductive and inductive reasoning are used to reach mathematical conclusions.	M2.1 Use simple logical reasoning to develop conclusions, recognizing that patterns and relationships present in the environment assist them in reaching these conclusions.	M2.1a Explain verbally, graphically, or in writing the reasoning used to develop mathematical conclusions	8, 31
			M2.1b Explain verbally, graphically, or in writing patterns and relationships observed in the physical and living environment	2, 30, 32, 43
	Key Idea 3: Critical thinking skills are used in the solution of mathematical problems.	M3.1 Explore and solve problems generated from school, home, and community situations, using concrete objects or manipulative materials when possible.	M3.1a Use appropriate scientific tools, such as metric rulers, spring scale, pan balance, graph paper, thermometers [Fahrenheit and Celsius], graduated cylinder to solve problems about the natural world	3, 7, 24, 27
STANDARD 1 Analysis, Inquiry, and Design **SCIENTIFIC INQUIRY (S)**	*Key Idea 1:* The central purpose of scientific inquiry is to develop explanations of natural phenomena in a continuing creative process.	S 1.1 Ask "why" questions in attempts to seek greater understanding concerning objects and events they have observed and heard about.	S1.1a Observe and discuss objects and events and record observations	1
			S1.1b Articulate appropriate questions based on observations	1
		S1.2 Question the explanations they hear from others and read about, seeking clarification and comparing them with their own observations and understandings.	S1.2a Identify similarities and differences between explanations received from others or in print and personal observations or understandings	1
		S1.3 Develop relationships among observations to construct descriptions of objects and events and to form their own tentative explanations of what they have observed.	S1.3a Clearly express a tentative explanation or description which can be tested	6

New York State Learning Standards Correlation Chart

Learning Standards	Key Ideas	Performance Indicators	Major Understandings	Coach Lesson(s)
STANDARD 1 Analysis, Inquiry, and Design **SCIENTIFIC INQUIRY (S)**	**Key Idea 2:** Beyond the use of reasoning and consensus, scientific inquiry involves the testing of proposed explanations involving the use of conventional techniques and procedures and usually requiring considerable ingenuity.	S2.1 Develop written plans for exploring phenomena or for evaluating explanations guided by questions or proposed explanations they have helped formulate.	S2.1a Indicate materials to be used and steps to follow to conduct the investigation and describe how data will be recorded (journal, dates and times, etc.)	6
		S2.2 Share their research plans with others and revise them based on their suggestions.	S2.2a Explain the steps of a plan to others, actively listening to their suggestions for possible modification of the plan, seeking clarification and understanding of the suggestions and modifying the plan where appropriate	6
		S2.3 Carry out their plans for exploring phenomena through direct observation and through the use of simple instruments that permit measurement of quantities, such as length, volume, mass, temperature, and time.	S2.3a Use appropriate "inquiry and process skills" to collect data	2–9, 18, 25, 31, 42
			S2.3b Record observations accurately and concisely	7
	Key Idea 3: The observations made while testing proposed explanations, when analyzed using conventional and invented methods, provide new insights into phenomena.	S3.1 Organize observations and measurements of objects and events through classification and the preparation of simple charts and tables.	S3.1a Accurately transfer data from a science journal or notes to appropriate graphic organizer	7
		S3.2 Interpret organized observations and measurements, recognizing simple patterns, sequences, and relationships.	S3.2a State, orally and in writing, any inferences or generalizations indicated by the data collected	8
		S3.3 Share their findings with others and actively seek their interpretations and ideas.	S3.3a Explain their findings to others, and actively listen to suggestions for possible interpretations and ideas	8, 9
		S3.4 Adjust their explanations and understandings of objects and events based on their findings and new ideas.	S3.4a State, orally and in writing, any inferences or generalizations indicated by the data, with appropriate modifications of their original prediction/explanation	9
			S3.4b State, orally and in writing, any new questions that arise from their investigation	9

Learning Standards	Key Ideas	Performance Indicators	Major Understandings	*Coach* Lesson(s)
STANDARD 1 Analysis, Inquiry, and Design: **ENGINEERING DESIGN (T)**	*Key Idea 1:* Engineering design is an iterative process involving modeling and optimization (finding the best solution within given constraints); this process is used to develop technological solutions to problems within given constraints.	T1.1 Describe objects, imaginary or real, that might be modeled or made differently and suggest ways in which the objects can be changed, fixed, or improved.	T1.1a Identify a simple/common object which might be improved and state the purpose of the improvement	10, 11
			T1.1b Identify features of an object that help or hinder the performance of the object	10, 11
			T1.1c Suggest ways the object can be made differently, fixed, or improved within given constraints	12
		T1.2 Investigate prior solutions and ideas from books, magazines, family, friends, neighbors, and community members.	T1.2a Identify appropriate questions to ask about the design of an object	11
			T1.2b Identify the appropriate resources to use to find out about the design of an object	11
			T1.2c Describe prior designs of the object	11
		T1.3 Generate ideas for possible solutions, individually and through group activity; apply age-appropriate mathematics and science skills; evaluate the ideas and determine the best solution; and explain reasons for the choices.	T1.3a List possible solutions, applying age-appropriate math and science skills	12
			T1.3b Develop and apply criteria to evaluate possible solutions	12
			T1.3c Select a solution consistent with given constraints and explain why it was chosen	12
		T1.4 Plan and build, under supervision, a model of the solution, using familiar materials, processes, and hand tools.	T1.4a Create a grade-appropriate graphic or plan listing all materials needed, showing sizes of parts, indicating how things will fit together, and detailing steps for assembly	12
			T1.4b Build a model of the object, modifying the plan as necessary	12
		T1.5 Discuss how best to test the solution; perform the test under teacher supervision; record and portray results through numerical and graphic means; discuss orally why things worked or didn't work; and summarize results in writing, suggesting ways to make the solution better.	T1.5a Determine a way to test the finished solution or model	13
			T1.5b Perform the test and record the results, numerically and/or graphically	13
			T1.5c Analyze results and suggest how to improve the solution or model, using oral, graphic, or written formats	13

New York State Learning Standards Correlation Chart

Learning Standards	Key Ideas	Performance Indicators	Major Understandings	Coach Lesson(s)
STANDARD 2: Information Systems Students will access, generate, process, and transfer information using appropriate technologies.				
STANDARD 2 **Information Systems (IS)**	*Key Idea 1:* Information technology is used to retrieve, process, and communicate information and as a tool to enhance learning.	IS1	• use computer technology, traditional paper-based resources, and interpersonal discussions to learn, do, and share science in the classroom • select appropriate hardware and software that aids in word processing, creating databases, telecommunications, graphing, data display, and other tasks • use information technology to link the classroom to world events	7, 9
	Key Idea 2: Knowledge of the impacts and limitations of information systems is essential to its effectiveness and ethical use.	IS2	• use a variety of media to access scientific information • consult several sources of information and points of view before drawing conclusions • identify and report sources in oral and written communications	9, 12
	Key Idea 3: Information technology can have positive and negative impacts on society, depending upon how it is used.	IS3	• distinguish fact from fiction (presenting opinion as fact is contrary to the scientific process) • demonstrate an ability to critically evaluate information and misinformation • recognize the impact of information technology on the daily life of students	1, 8, 10

STANDARD 4 : The Physical Setting and Living Environment
Students will understand and apply scientific concepts, principles, and theories pertaining to the physical setting and living environment, and recognize the historical development of ideas in science.

STANDARD 4 The Physical Setting (PS)	Key Idea 1: The Earth and celestial phenomena can be described by principles of relative motion and perspective.	PS1.1 Describe patterns of daily, monthly, and seasonal changes in their environment.	PS1.1a Natural cycles and patterns include: • Earth spinning around once every 24 hours (rotation), resulting in day and night • Earth moving in a path around the Sun (revolution), resulting in one Earth year • the length of daylight and darkness varying with the seasons • weather changing from day to day and through the seasons • the appearance of the Moon changing as it moves in a path around Earth to complete a single cycle	14, 15, 16
			PS1.1b Humans organize time into units based on natural motions of Earth: • second, minute, hour • week, month	17
			PS1.1c The Sun and other stars appear to move in a recognizable pattern both daily and seasonally.	2, 14
	Key Idea 2: Many of the phenomena that we observe on Earth involve interactions among components of air, water, and land.	PS2.1 Describe the relationship among air, water, and land on Earth.	PS2.1a Weather is the condition of the outside air at a particular moment.	18
			PS2.1b Weather can be described and measured by: • temperature • wind speed and direction • form and amount of precipitation • general sky conditions (cloudy, sunny, partly cloudy)	18
			PS2.1c Water is recycled by natural processes on Earth. • evaporation: changing of water (liquid) into water vapor (gas) • condensation: changing of water vapor (gas) into water (liquid) • precipitation: rain, sleet, snow, hail • runoff: water flowing on Earth's surface • groundwater: water that moves downward into the ground	19, 23
			PS2.1d Erosion and deposition result from the interaction among air, water, and land. • interaction between air and water breaks down earth materials • pieces of earth material may be moved by air, water, wind, and gravity • pieces of earth material will settle or deposit on land or in the water in different places • soil is composed of broken-down pieces of living and nonliving earth material	20
			PS2.1e Extreme natural events (floods, earthquakes, fires, volcanic eruptions, hurricanes, tornadoes, and other severe storms) may have positive or negative impacts on living things.	21

New York State Learning Standards Correlation Chart

Learning Standards	Key Ideas	Performance Indicators	Major Understandings	*Coach* Lesson(s)
STANDARD 4 **The Physical Setting (PS)**	*Key Idea 3:* Matter is made up of particles whose properties determine the observable characteristics of matter and its reactivity.	PS3.1 Observe and describe properties of materials, using appropriate tools.	PS3.1a Matter takes up space and has mass. Two objects cannot occupy the same place at the same time.	22
			PS3.1b Matter has properties (color, hardness, odor, sound, taste, etc.) that can be observed through the senses.	22
			PS3.1c Objects have properties that can be observed, described, and/or measured: length, width, volume, size, shape, mass or weight, temperature, texture, flexibility, reflectiveness of light.	22
			PS3.1d Measurements can be made with standard metric units and nonstandard units.	4
			PS3.1e The material(s) an object is made up of determine some specific properties of the object (sink/float, conductivity, magnetism). Properties can be observed or measured with tools such as hand lenses, thermometers, metric rulers, balances, magnets, circuit testers, and graduated cylinders.	3, 22, 24
			PS3.1f Objects and/or materials can be sorted or classified according to their properties.	25
			PS3.1g Some properties of an object are dependent on the conditions of the present surroundings in which the object exists. For example: • temperature - hot or cold • lighting - shadows, color • moisture - wet or dry	28
		PS3.2 Describe chemical and physical changes, including changes in states of matter.	PS3.2a Matter exists in three states: solid, liquid, gas. • solids have a definite shape and volume • liquids do not have a definite shape but have a definite volume • gases do not hold their shape or volume	22, 23
			PS3.2b Temperature can affect the state of matter of a substance.	23
			PS3.2c Changes in the properties or materials of objects can be observed and described.	23

Learning Standards	Key Ideas	Performance Indicators	Major Understandings	Coach Lesson(s)
STANDARD 4 **The Physical Setting (PS)**	*Key Idea 4:* Energy exists in many forms, and when these forms change, energy is conserved.	PS4.1 Describe a variety of forms of energy (e.g., heat, chemical, light) and the changes that occur in objects when they interact with those forms of energy.	PS4.1a Energy exists in various forms: heat, electric, sound, chemical, mechanical, light.	26–29
			PS4.1b Energy can be transferred from one place to another.	26–29
			PS4.1c Some materials transfer energy better than others (heat and electricity).	27, 29
			PS4.1d Energy and matter interact: water is evaporated by the Sun's heat; a bulb is lighted by means of electrical current; a musical instrument is played to produce sound; dark colors may absorb light, light colors may reflect light.	28, 30
			PS4.1e Electricity travels in a closed circuit.	29
			PS4.1f Heat can be released in many ways, for example, by burning, rubbing (friction), or combining one substance with another.	27
			PS4.1g Interactions with forms of energy can be either helpful or harmful.	26–29
		PS4.2 Observe the way one form of energy can be transferred into another form of energy present in common situations (e.g., mechanical to heat energy, mechanical to electrical energy, chemical to heat energy).	PS4.2a Everyday events involve one form of energy being changed to another. • animals convert food to heat and motion • the Sun's energy warms the air and water	30
			PS4.2b Humans utilize interactions between matter and energy. • chemical to electrical, light, and heat: battery and bulb • electrical to sound (e.g., doorbell buzzer) • mechanical to sound (e.g., musical instruments, clapping) • light to electrical (e.g., solar-powered calculator)	30
	Key Idea 5: Energy and matter interact through forces that result in changes in motion.	PS5.1 Describe the effects of common forces (pushes and pulls) of objects, such as those caused by gravity, magnetism, and mechanical forces.	PS5.1a The position of an object can be described by locating it relative to another object or the background (e.g., on top of, next to, over, under, etc.).	31
			PS5.1b The position or direction of motion of an object can be changed by pushing or pulling.	31, 32
			PS5.1c The force of gravity pulls objects toward the center of Earth.	32
			PS5.1d The amount of change in the motion of an object is affected by friction.	32, 34
			PS5.1e Magnetism is a force that may attract or repel certain materials.	33
			PS5.1f Mechanical energy may cause change in motion through the application of force and through the use of simple machines such as pulleys, levers, and inclined planes.	34
		PS5.2 Describe how forces can operate across distances.	PS5.2a The forces of gravity and magnetism can affect objects through gases, liquids, and solids.	32, 33
			PS5.2b The force of magnetism on objects decreases as distance increases.	33

New York State Learning Standards Correlation Chart

Learning Standards	Key Ideas	Performance Indicators	Major Understandings	Coach Lesson(s)
STANDARD 4 **The Living Environment (LE)**	*Key Idea 1:* Living things are both similar to and different from each other and from nonliving things.	LE1.1 Describe the characteristics of and variations between living and nonliving things.	LE1.1a Animals need air, water, and food in order to live and thrive.	35
			LE1.1b Plants require air, water, nutrients, and light in order to live and thrive.	35
			LE1.1c Nonliving things do not live and thrive.	35
			LE1.1d Nonliving things can be human-created or naturally occurring.	35
		LE1.2 Describe the life processes common to all living things.	LE1.2a Living things grow, take in nutrients, breathe, reproduce, eliminate waste, and die.	35
	Key Idea 2: Organisms inherit genetic information in a variety of ways that result in continuity of structure and function between parents and offspring.	LE2.1 Recognize that traits of living things are both inherited and acquired or learned.	LE2.1a Some traits of living things have been inherited (e.g., color of flowers and number of limbs of animals).	36
			LE2.1b Some characteristics result from an individual's interactions with the environment and cannot be inherited by the next generation (e.g., having scars; riding a bicycle).	36
		LE2.2 Recognize that for humans and other living things there is genetic continuity between generations.	LE2.2a Plants and animals closely resemble their parents and other individuals in their species.	36
			LE2.2b Plants and animals can transfer specific traits to their offspring when they reproduce.	36
	Key Idea 3: Individual organisms and species change over time.	LE3.1 Describe how the structures of plants and animals complement the environment of the plant or animal.	LE3.1a Each animal has different structures that serve different functions in growth, survival, and reproduction. • wings, legs, or fins enable some animals to seek shelter and escape predators • the mouth, including teeth, jaws, and tongue, enables some animals to eat and drink • eyes, nose, ears, tongue, and skin of some animals enable the animals to sense their surroundings • claws, shells, spines, feathers, fur, scales, and color of body covering enable some animals to protect themselves from predators and other environmental conditions, or enable them to obtain food • some animals have parts that are used to produce sounds and smells to help the animal meet its needs • the characteristics of some animals change as seasonal conditions change (e.g., fur grows and is shed to help regulate body heat; body fat is a form of stored energy and it changes as the seasons change)	39

Learning Standards	Key Ideas	Performance Indicators	Major Understandings	Coach Lesson(s)
STANDARD 4 **The Living Environment (LE)**			**LE3.1b** Each plant has different structures that serve different functions in growth, survival, and reproduction. • roots help support the plant and take in water and nutrients • leaves help plants utilize sunlight to make food for the plant • stems, stalks, trunks, and other similar structures provide support for the plant • some plants have flowers • flowers are reproductive structures of plants that produce fruit which contains seeds • seeds contain stored food that aids in germination and the growth of young plants	37
			LE3.1c In order to survive in their environment, plants and animals must be adapted to that environment. • seeds disperse by a plant's own mechanism and/or in a variety of ways that can include wind, water, and animals • leaf, flower, stem, and root adaptations may include variations in size, shape, thickness, color, smell, and texture • animal adaptations include coloration for warning or attraction, camouflage, defense mechanisms, movement, hibernation, and migration	37, 39
		LE3.2 Observe that differences within a species may give individuals an advantage in surviving and reproducing.	**LE3.2a** Individuals within a species may compete with each other for food, mates, space, water, and shelter in their environment.	43
			LE3.2b All individuals have variations, and because of these variations, individuals of a species may have an advantage in surviving and reproducing.	43

New York State Learning Standards Correlation Chart

Learning Standards	Key Ideas	Performance Indicators	Major Understandings	Coach Lesson(s)
STANDARD 4 **The Living Environment (LE)**	*Key Idea 4:* The continuity of life is sustained through reproduction and development.	LE4.1 Describe the major stages in the life cycles of selected plants and animals.	LE4.1a Plants and animals have life cycles. These may include beginning of a life, development into an adult, reproduction as an adult, and eventually death.	38, 40
			LE4.1b Each kind of plant goes through its own stages of growth and development that may include seed, young plant, and mature plant.	38
			LE4.1c The length of time from beginning of development to death of the plant is called its life span.	38
			LE4.1d Life cycles of some plants include changes from seed to mature plant.	38
			LE4.1e Each generation of animals goes through changes in form from young to adult. This completed sequence of changes in form is called a life cycle. Some insects change from egg to larva to pupa to adult.	40
			LE4.1f Each kind of animal goes through its own stages of growth and development during its life span.	40
			LE4.1g The length of time from an animal's birth to its death is called its life span. Life spans of different animals vary.	40
		LE4.2 Describe evidence of growth, repair, and maintenance, such as nails, hair, and bone, and the healing of cuts and bruises.	LE4.2a Growth is the process by which plants and animals increase in size.	35
			LE4.2b Food supplies the energy and materials necessary for growth and repair.	35
	Key Idea 5: Organisms maintain a dynamic equilibrium that sustains life.	LE5.1 Describe basic life functions of common living specimens (e.g., guppies, mealworms, gerbils).	LE5.1a All living things grow, take in nutrients, breathe, reproduce, and eliminate waste.	35
			LE5.1b An organism's external physical features can enable it to carry out life functions in its particular environment.	37, 39

Learning Standards	Key Ideas	Performance Indicators	Major Understandings	*Coach* Lesson(s)
STANDARD 4 **The Living Environment (LE)**		LE5.2: Describe some survival behaviors of common living specimens.	LE5.2a Plants respond to changes in their environment. For example, the leaves of some green plants change position as the direction of light changes; the parts of some plants undergo seasonal changes that enable the plant to grow; seeds germinate, and leaves form and grow.	35
			LE5.2b Animals respond to change in their environment, (e.g., perspiration, heart rate, breathing rate, eye blinking, shivering, and salivating).	39
			LE5.2c Senses can provide essential information (regarding danger, food, mates, etc.) to animals about their environment.	39
			LE5.2d Some animals, including humans, move from place to place to meet their needs.	39
			LE5.2e Particular animal characteristics are influenced by changing environmental conditions including: fat storage in winter, coat thickness in winter, camouflage, shedding of fur.	39
			LE5.2f Some animal behaviors are influenced by environmental conditions. These behaviors may include: nest building, hibernating, hunting, migrating, and communicating.	36
			LE5.2g The health, growth, and development of organisms are affected by environmental conditions such as the availability of food, air, water, space, shelter, heat, and sunlight.	43
		LE5.3 Describe the factors that help promote good health and growth in humans.	LE5.3a Humans need a variety of healthy foods, exercise, and rest in order to grow and maintain good health.	41
			LE5.3b Good health habits include hand washing and personal cleanliness; avoiding harmful substances (including alcohol, tobacco, illicit drugs); eating a balanced diet; engaging in regular exercise.	41

New York State Learning Standards Correlation Chart

Learning Standards	Key Ideas	Performance Indicators	Major Understandings	Coach Lesson(s)
STANDARD 4 **The Living Environment (LE)**	*Key Idea 6:* Plants and animals depend on each other and their physical environment.	LE6.1 Describe how plants and animals, including humans, depend upon each other and the nonliving environment.	LE6.1a Green plants are producers because they provide the basic food supply for themselves and animals.	42
			LE6.1b All animals depend on plants. Some animals (predators) eat other animals (prey).	42, 43
			LE6.1c Animals that eat plants for food may in turn become food for other animals. This sequence is called a food chain.	42
			LE6.1d Decomposers are living things that play a vital role in recycling nutrients.	42
			LE6.1e An organism's pattern of behavior is related to the nature of that organism's environment, including the kinds and numbers of other organisms present, the availability of food and other resources, and the physical characteristics of the environment.	36, 43
			LE6.1f When the environment changes, some plants and animals survive and reproduce, and others die or move to new locations.	43
		LE6.2 Describe the relationship of the Sun as an energy source for living and nonliving cycles.	LE6.2a Plants manufacture food by utilizing air, water, and energy from the Sun.	42
			LE6.2b The Sun's energy is transferred on Earth from plants to animals through the food chain.	42
			LE6.2c Heat energy from the Sun powers the water cycle (see Physical Science Key Idea 2).	19
	Key Idea 7: Human decisions and activities have had a profound impact on the physical and living environment.	LE7.1 Identify ways in which humans have changed their environment and the effects of those changes.	LE7.1a Humans depend on their natural and constructed environments.	44
			LE7.1b Over time humans have changed their environment by cultivating crops and raising animals, creating shelter, using energy, manufacturing goods, developing means of transportation, changing populations, and carrying out other activities.	44
			LE7.1c Humans, as individuals or communities, change environments in ways that can be either helpful or harmful for themselves and other organisms.	44

Learning Standards	Key Ideas	Performance Indicators	Major Understandings	Coach Lesson(s)
Standard 6: Interconnectedness: Common Themes Students will understand the relationships and common themes that connect mathematics, science, and technology and apply the themes to these and other areas of learning.					
STANDARD 6: Interconnectedness: Common Themes (CT)	*Key Idea 1:* Through systems thinking, people can recognize the commonalities that exist among all systems and how parts of a system interrelate and combine to perform specific functions.	CT1	• observe and describe interactions among components of simple systems • identify common things that can be considered to be systems (e.g., a plant, a transportation system, human beings)	2, 37, 39
	Key Idea 2: Models are simplified representations of objects, structures, or systems, used in analysis, explanation, or design.	CT2	• analyze, construct, and operate models in order to discover attributes of the real thing • discover that a model of something is different from the real thing but can be used to study the real thing • use different types of models, such as graphs, sketches, diagrams, and maps, to represent various aspects of the real world	7, 12, 13
	Key Idea 3: The grouping of magnitudes of size, time, frequency, and pressures or other units of measurement into a series of relative order provides a useful way to deal with the immense range and the changes in scale that affect behavior and design of systems.	CT3	• observe that things in nature and things that people make have very different sizes, weights, and ages • recognize that almost anything has limits on how big or small it can be	3
	Key Idea 4: Equilibrium is a state of stability due either to a lack of changes (static equilibrium) or a balance between opposing forces (dynamic equilibrium).	CT4	• observe that things change in some ways and stay the same in some ways • recognize that things can change in different ways such as size, weight, color, and movement. Some small changes can be detected by taking measurements.	2, 3, 22
	Key Idea 5: Identifying patterns of change is necessary for making predictions about future behavior and conditions.	CT5	• use simple instruments to measure such quantities as distance, size, and weight and look for patterns in the data • analyze data by making tables and graphs and looking for patterns of change	2, 3, 7
	Key Idea 6: In order to arrive at the best solution that meets criteria within constraints, it is often necessary to make trade-offs.	CT6	• choose the best alternative of a set of solutions under given constraints • explain the criteria used in selecting a solution orally and in writing	8, 13

New York State Learning Standards Correlation Chart

Learning Standards	Key Ideas	Performance Indicators	Major Understandings	Coach Lesson(s)
Standard 7: Interdisciplinary Problem Solving Students will understand the relationships and common themes that connect mathematics, science, and technology and apply the themes to these and other areas of learning.					
STANDARD 7 Interdisciplinary Problem Solving (IP)	*Key Idea 1:* The knowledge and skills of mathematics, science, and technology are used together to make informed decisions and solve problems, especially those relating to issues of science/technology/society, consumer decision making, design, and inquiry into phenomena.	IP1	• analyze science/technology/society problems and issues that affect their home, school, or community, and carry out a remedial course of action • make informed consumer decisions by applying knowledge about the attributes of particular products and making cost/benefit trade-offs to arrive at an optimal choice • design solutions to problems involving a familiar and real context, investigate related science concepts to determine the solution, and use mathematics to model, quantify, measure, and compute • observe phenomena and evaluate them scientifically and mathematically by conducting a fair test of the effect of variables and using mathematical knowledge and technological tools to collect, analyze, and present data and conclusions	11, 12, 13
	Key Idea 2: Solving interdisciplinary problems involves a variety of skills and strategies, including effective work habits; gathering and processing information; generating and analyzing ideas; realizing ideas; making connections among the common themes of mathematics, science, and technology; and presenting results.	IP2	• work effectively • gather and process information • generate and analyze ideas • observe common themes • realize ideas • present results	6, 12

New York State Core Curriculum Science Skills

GENERAL SKILLS IN SCIENCE
Science is an ongoing process. Most often there is a question or problem that initiates an investigation searching for a possible solution or solutions. There is no single prescribed scientific method to govern an investigation. It is important that students practice the skills outlined below.
General Skills
GSi. follow safety procedures in the classroom, laboratory, and field
GSii. safely and accurately use the following tools: • hand lens • gram weights • measuring cups • ruler (metric) • spring scale • graduated cylinder • balance • thermometer (°C, °F) • timepiece(s)
GSiii. develop an appreciation of and respect for all learning environments (classroom, laboratory, field, etc.)
GSiv. manipulate materials through teacher direction and free discovery
GSv. use information systems appropriately
GSvi. select appropriate standard and nonstandard measurement tools for measurement activities
GSvii. estimate, find, and communicate measurements, using standard and nonstandard units
GSviii. use and record appropriate units for measured or calculated values
GSix. order and sequence objects and/or events
GSx. classify objects according to an established scheme
GSxi. generate a scheme for classification
GSxii. utilize senses optimally for making observations
GSxiii. observe, analyze, and report observations of objects and events
GSxiv. observe, identify, and communicate patterns
GSxv. observe, identify, and communicate cause-and-effect relationships
GSxvi. generate appropriate questions (teacher and student based) in response to observations, events, and other experiences
GSxvii. observe, collect, organize, and appropriately record data, then accurately interpret results
GSxviii. collect and organize data, choosing the appropriate representation: • journal entries • graphic representations • drawings/pictorial representations
GSxix. make predictions based on prior experiences and/or information
GSxx. compare and contrast organisms/objects/events in the living and physical environments
GSxxi. identify and control variables/factors
GSxxii. plan, design, and implement a short-term and long-term investigation based on a student- or teacher-posed problem
GSxxiii. communicate procedures and conclusions through oral and written presentations

CHAPTER 1: Scientific Inquiry

Lesson 1: Observe and Ask Questions

S1.1a, b; S1.2a, IS3

Getting the Idea

Key Words
observe
senses
measure
fact
opinion

Scientists ask many questions about the world around them. Then they try to find answers. Think about questions you might ask. Do you ever wonder about what you see or hear? Do you want to know why some things happen? Do you ask how objects work or what things are made of? If so, you are already doing what scientists do.

Observing

When you **observe**, you use your senses to learn more about something. Your **senses** are how you see, hear, touch, smell, or taste. Observing helps you answer questions. Suppose you saw crumbs on the kitchen counter near the toaster.

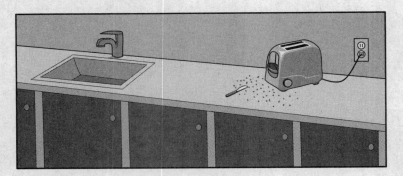

You might wonder what someone had for a snack. You might ask these questions:

- Is there any jam, butter, or icing around?
- Is there a knife nearby that was used with a spread?
- Is there a pastry wrapper in the garbage can?
- Are the crumbs on the counter from bread?

Lesson 1: Observe and Ask Questions

You can answer these questions by observing. You may notice the smell of toast in the air. You may touch the crumbs to see if they are crunchy or soft. You could even taste the crumbs to see if they taste sweet, like a toasted pastry. The answers might help you figure out what the snack was. When you figure something out in this way, you act like a scientist.

Scientists ask many questions about the world around them. Then they try to find answers. Think about questions you might ask. Do you ever wonder about what you see or hear? Do you want to know why some things happen? Do you ask how objects work or what things are made of? If so, you already think like a scientist.

Ask Questions

Suppose you observe a plant that grows near your home. You wonder what a plant needs to grow from a seed. You know that a plant needs water. Does it need sunlight, too? If so, how much?

Scientists ask questions they can answer by doing investigations. When you investigate, you study something carefully to answer a question about it. You may read a book about plants to figure out the answer. You may also design a setup like the one shown below. You will learn more about these kinds of investigations in Chapter 2.

There are many ways to investigate. For example, you might ask how many books fit on your shelf. You can test this question yourself. First, fill the shelf with books. Then count the books. Now you have the answer.

> **Did You Know?**
>
> Albert Einstein was one of the world's greatest scientists. He said, "The important thing is not to stop questioning."

Scientists base the answers to their questions on observations and measurements. To **measure** is to find the size or amount of something. Scientists write down their observations and measurements. It is important to keep a good record of all your work. Then, you can show others what you figured out. Others can also try what you did to see if they get the same answers. This is how scientists learn from one another.

Questions Based on Facts

Suppose you ask, "Which is better, strawberry jelly or grape jelly?" Science cannot answer this kind of question. You cannot test or measure if something is "better." You could ask which is sweeter or which costs more. You can test those questions.

In science, questions must be about facts. A **fact** is a piece of information that is true. It is supported by observations. You can make an observation or repeat an investigation to check a fact. A fact does not change from one person or place to the next. For example, it is a fact that plants need air and water to survive. Wherever you go on Earth, you will find that plants need air and water. You can make observations that support this fact.

A statement of fact can be proved true or false. Here are some examples of statements of fact:

- This rock weighs 30 grams.
- Water freezes at 0°C.
- The plant grew 1 inch.
- Rain comes from clouds.

Not all statements of fact are true. You can check statements of fact to see if they are true. One way to check is to use reference materials, such as books or the Internet. The reference materials must be from a source you can trust. Encyclopedias, government Web sites, and science journals are sources you can trust. You can check some facts by measuring. You can weigh an object, find its volume, or measure its length to check if a statement is true.

Questions Based on Opinions

An **opinion** is a person's view of something. Unlike a fact, an opinion cannot be proved true or false. An opinion is a belief, a feeling, or a judgment. Like the question about which jelly is better, you cannot test an opinion in science.

You may think New York is the most beautiful state in the country. However, your friend in Hawaii may think that state is the most beautiful. What you think or what your friend thinks is an opinion.

Certain words are clues that a statement is an opinion. Look for phrases such as *I think*, *I feel*, or *I believe*. Words such as *worst*, *best*, and *should* also show that a statement is an opinion. Here are some examples of statements of opinion.

- I think that book is very interesting.
- You should read that magazine.
- That is the worst movie I ever saw.
- I believe there is life on other planets.

You can give reasons why you hold your opinion. You can explain why you do not agree with someone else's opinion. But you cannot prove that an opinion is true or false. Knowing the difference between fact and opinion is very important. Scientific knowledge is not based on opinions.

DISCUSSION QUESTION

Name one fact and one opinion about today's weather. Use an observation to support the fact.

LESSON REVIEW

1. What do you ALWAYS use when you observe something?
   - A. a notebook
   - B. a measuring tool
   - C. your imagination
   - D. your senses

2. A student wants to know how baby robins learn to fly. She observes a family of robins outside her window. She records what she sees in her science journal. Which things do NOT belong in a science journal?

   A. written notes about the robins
   B. drawings of the robins
   C. a made-up story about a talking robin
   D. photos of the robins

3. Which sentence about statements of fact is TRUE?

   A. They cannot be proved true.
   B. They can be proved true or false.
   C. They are always true.
   D. They are always false.

4. Which one is a statement of opinion?

   A. The rock has a mass of 20 g.
   B. Venus is the second planet from the sun.
   C. Ocean waves sound beautiful.
   D. Two inches of rain fell on April 12.

5. Which of these is NOT a reliable source for checking a statement of fact?

   A. a government Web site
   B. an encyclopedia
   C. a science journal
   D. a movie review

Lesson 2: Identify Systems and Patterns

M2.1b; S2.3a; PS1.1c; CT1, 4, 5

Getting the Idea

Key Words
pattern
sequence
system
trend
prediction

People often try to find patterns in nature. A **pattern** is something that repeats, such as a set of events. Patterns help you understand the natural world. Patterns also help you know what to expect in the future. Understanding natural patterns is an important part of thinking like a scientist.

Patterns of Change

Scientists observe and look for patterns in nature. The sun rises in the east every morning, and it is day. Every evening the sun sets in the west, and it is night. Here in New York State, this pattern of sunrise and sunset repeats every 24 hours. You can measure time in days by observing the motions of the sun.

You can observe patterns in the night sky, too. One pattern is the way that the lit part of the moon seems to change shape. Some of these changes, called phases, are shown below. The phases always happen in the same order. The phases form a **sequence**, a pattern of events that always happen in the same order.

Four Main Phases of the Moon

New moon   First quarter   Full moon   Last quarter

The patterns you can observe in the sun and the moon repeat because of the positions of Earth, the sun, and the moon. Earth, the sun, and the moon are part of a **system**, a group of parts that work together.

Did You Know?

New York's state insect is the ladybug. Ladybugs hibernate in winter. They huddle together under leaves or grasses and do not move. The ladybugs become active again when the temperature reaches about 59°F (15°C).

The change of seasons also forms a pattern. In New York, seeds sprout in spring. Temperatures get warmer in summer. Leaves change color and fall in autumn. Many animals slow down in winter. The four seasons happen in the same sequence year after year.

You can observe patterns in the way many living things grow. As trees grow, the wood forms a pattern of rings. You can see these rings inside the trunk of a tree.

A tree grows new rings each year just under the bark. Each ring is made of two sections. First, a light section of the ring grows in spring. Then, in summer and fall, a darker section grows.

Trends

Third graders are often taller than second graders. Fourth graders are often taller than third graders. You know that as children get older, they get taller. This is a **trend**, which is a change that happens in a steady way. The graph below shows how the heights of some students changed over ten years. You can see the trend displayed on the graph. You will learn more about making and reading graphs in Lesson 7.

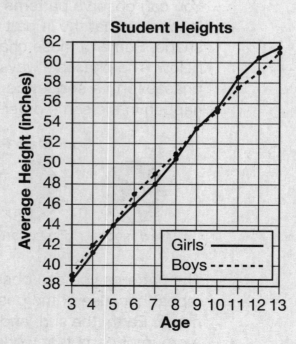

Use Patterns to Make Predictions

Because patterns repeat themselves, you can use them to tell what will happen next. A **prediction** is a guess about what is likely to happen in the future. Suppose you notice a pattern in the weather. Whenever a certain type of large, dark cloud fills the sky, you observe that it rains. The next time you see those dark clouds, you predict that it will rain again.

You can look at different kinds of weather maps and guess where a storm might hit next. You can look at the path of the storm and make good guesses. The map below shows the path of a hurricane.

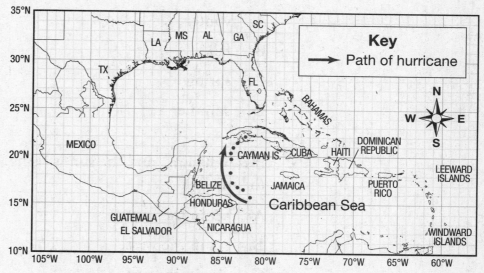

The map shows that a hurricane is moving northward over the Caribbean Sea. You might predict that the storm will hit southern Florida.

Scientists observe many different kinds of patterns to make predictions. Predictions can help people learn more and can also help keep people safe.

DISCUSSION QUESTION

What are two different patterns that affect your life? Identify and describe them.

LESSON REVIEW

1. Which statement about patterns is TRUE?

    A. They are things that stay the same.

    B. They always repeat one time each year.

    C. Scientists find them by making observations.

    D. They are not found in nature.

2. A friend shows you a photo of her upstate New York home. The photo shows red leaves falling from a tree. In which season was the photo MOST LIKELY taken?

    A. winter

    B. fall

    C. spring

    D. summer

3. It is 6:30 a.m., and the sun is rising. In 24 hours, the sun will

    A. set in the east.

    B. set in the west.

    C. rise in the east.

    D. rise in the west.

Lesson 3: Tools for Observing and Measuring

M3.1a; S2.3a; PS3.1e, CT3, 4, 5

Getting the Idea

When you investigate something in science, you will often use certain tools or instruments to make observations. You will also use many different tools to measure things. This lesson will introduce you to some science tools. You will learn which tools are best for the types of measurements and observations you want to make.

Key Words
- mass
- pan balance
- weight
- spring scale
- volume
- graduated cylinder
- thermometer

Use Tools to Observe

As you carry out investigations, you might need to look at something very closely. A simple hand lens will help you see small objects. It will also help you see details in large objects.

Scientists also use tools such as cameras to record events. You might be able to use a video camera in your own investigations. For example, you might want to record a moth coming out of its cocoon. Or you might want to record how different people work to solve the same puzzle.

Duplicating any part of this book is prohibited by law.

Use Tools to Measure

Scientists use measuring tools to collect information. This information can help them answer questions. Scientists must choose the correct tool to make a measurement. You would not use a ruler to measure the distance from your home to your school. A ruler is too small to measure a large distance. The measuring tool should be the right size for the object you need to measure.

Measuring Size

You can measure length, width, height, and depth with a meterstick or ruler. Make sure the end of the stick, or the 0 mark, is exactly at one edge of an object. If an object is round or curved, you can use a measuring tape. A measuring tape bends so it can wrap around an object.

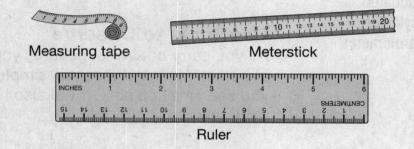

Measuring tape   Meterstick

Ruler

Measuring Weight and Mass

Mass is the amount of matter that makes up an object. You can measure mass using a **pan balance**. A pan balance compares the masses of things placed on the two pans. **Weight** is the measure of the pull of gravity on an object. A **spring scale** measures weight. To be accurate, you must wait until the numbers stop moving before you record the weight.

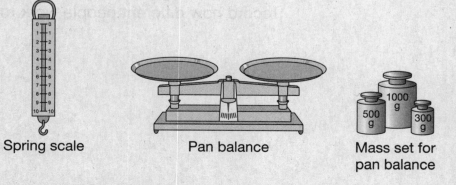

Spring scale   Pan balance   Mass set for pan balance

Lesson 3: **Tools for Observing and Measuring**

Measuring Volume

Volume is the amount of space that something takes up. You can measure volume with a measuring cup, measuring spoons, a beaker, or a **graduated cylinder**. Sometimes you measure the volume of certain materials before you mix them. Sometimes you measure volume before and after an investigation to determine how it changed. When you measure volume, make sure that your eyes are at the same level as the marks on the cup or cylinder. This helps you read the volume accurately.

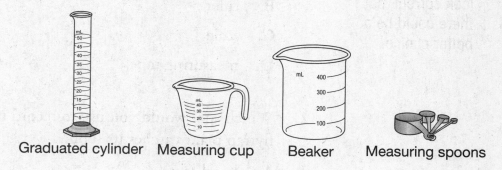

Graduated cylinder   Measuring cup   Beaker   Measuring spoons

When you measure a liquid with a teaspoon or tablespoon, make sure the liquid is even with the top edge of the spoon. When measuring powder, use a ruler to push off any extra powder. The powder should be even with the top of the spoon.

Measuring Time

A clock or a watch with a second hand measures time. A stopwatch is a tool that you can start and stop to find out how long it takes something to happen. You will learn more about measuring time in Lesson 17.

Measuring Temperature

A **thermometer** tells you how warm something is. You might use a thermometer to measure air temperature. You can measure the temperature of many different objects and substances. You might also use a thermometer to find out how much the temperature changes over time. You will learn more about measuring temperature in Lesson 18.

Test Tips...

Do not pick an answer too quickly. Be sure to look at all the answers before you decide. One answer might look correct, but there could be a better choice.

DISCUSSION QUESTION

When you choose a tool for measuring something, what are two things you should consider?

LESSON REVIEW

1. Which tool should you use to measure the distance around a ball?

    A. graduated cylinder

    B. ruler

    C. scale

    D. measuring tape

2. Which tool would you use to record how the Venus flytrap plant catches its food?

    A. hand lens

    B. scale

    C. telescope

    D. video camera

3. Which example is it BEST to measure with a stopwatch?

    A. the distance between your house and school

    B. the time it takes a toy car to roll down a small ramp

    C. how many hours an eclipse lasts

    D. how far a car can go on a gallon of gasoline

Lesson 4: Measuring Systems

M1.1a; S2.3a; PS3.1d

Getting the Idea

How wide is your desk? You might not be sure how to answer that. To get an idea, you see how many times your pencil fits across your desk. You figure out that your desk is 5 pencils long. You tell your friend your answer, but your friend says you are wrong. She says your desk is 4 pencils long. Another student might say your desk is 6 pencils long. You can't all be right. Or can you? In this lesson, you will learn what it means to measure. What you use to measure is actually a very important choice.

Key Words
- measure
- unit of measurement
- metric system
- customary unit

Measuring

Recall that to **measure** is to find out the amount or size of something. To do that, you compare a known unit to an object or action. Suppose a student records the mass of an apple as 600. Is this information helpful to you? It does not help you because the information does not include a unit of measurement.

A **unit of measurement** is a standard amount. For instance, a meter is a unit of length. One meter is always the same length, no matter what. Every measurement must have a number and a unit. Take a look at how two different students measured the same desk using their pencils.

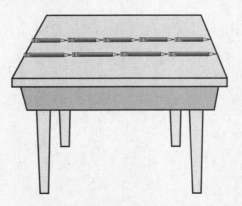

If you say your desk is 5 pencils long, your unit of measurement is a pencil. The problem with this unit is that it is not standard. Pencils come in different sizes. The student who said your desk is 4 pencils long also used a pencil. But the student used a different pencil. He even measured your desk and showed you. Pencils come in different sizes. That is why the answers were not the same.

People based many early units of measurements on parts of the human body. A foot, for example, was the length of a grown man's foot. But like the pencil, the length of the human foot is not standard. It can vary from person to person. People realized they needed to have units of measurement that are always the same. Then, you can compare measurements fairly.

The Metric System

In the late 1700s, a group of scientists developed the system of measurement known as the **metric system**. In the metric system, each unit has a standard value that does not vary. When people use the same system of measurement, it is much easier to share information.

Mass is the amount of matter in an object. The base unit of mass is the gram. Volume is the amount of space something takes up. The base unit of volume is the liter.

The metric system is based on the number 10. For each kind of measurement, there is a base unit. Other units relate to the base unit by a factor of 10. For example, the unit of length is the meter. A kilometer is 1000 meters. A centimeter is $\frac{1}{100}$ of a meter.

Lesson 4: Measuring Systems

The table below shows some base units and some other units that are often used. Notice that the same word parts are used for the same factors of ten. For example, any base unit multiplied by 1000 has the word part *kilo-* in front of it.

Metric Units

Measurement	Base Unit	Other Units You May Use
Length	meter	• 1000 meters = 1 kilometer • $\frac{1}{100}$ meter = 1 centimeter • $\frac{1}{1000}$ meter = 1 millimeter
Mass	gram	• 1000 grams = 1 kilogram • $\frac{1}{100}$ gram = 1 centigram • $\frac{1}{1000}$ gram = 1 milligram
Volume	liter	• $\frac{1}{1000}$ liter = 1 milliliter

Customary Units

People in most countries use the metric system. But in the United States, we do not always use the metric system. We use **customary units**, such as the foot, inch, gallon, and pound.

Customary Units

Measurement	Base Unit	Other Units You May Use
Length	foot	• 12 inches = 1 foot • 3 feet = 1 yard
Weight	pound	• 16 ounces = 1 pound
Volume	quart	• 2 pints = 1 quart • 4 quarts = 1 gallon

Customary units do not use factors of ten. Parts of the words do not show how units relate. These things make customary units harder to remember. People in the United States use customary units in everyday life. However, scientists in the United States use the metric system.

Both customary and metric units are standard. No matter where you are in the world, measurements such as 3 inches, 600 grams, or 2 liters always refer to the same amount of whatever it is they measure. They always mean the same thing.

DISCUSSION QUESTION

Scientists in all parts of the world use the same units to measure length, weight, and other things. What do you think might happen if they each used different units?

LESSON REVIEW

1. A scientist measures the height of a tall tree. Which unit does the scientist MOST LIKELY use?

   A. liter
   B. kilogram
   C. meter
   D. centimeter

2. Which one is correct?

   A. 1 gram = 1000 kilograms
   B. 1 centimeter = $\frac{1}{100}$ meter
   C. 1 milliliter = 100 liters
   D. 1 kilometer = $\frac{1}{100}$ meters

3. Which one is a customary unit?

   A. foot
   B. meter
   C. gram
   D. millimeter

4. You want to measure the distance from one side of your classroom to the other. Which is a standard unit of measurement for that distance?

   A. minute
   B. foot
   C. kilometer
   D. pound

Lesson 5: Safety in Science

S2.3a

Getting the Idea

Key Words
safety goggles
safety symbol

When you ride a bike, you follow safety rules. They keep you from getting hurt. You must also follow safety rules when you use science tools and materials. Following safety rules is part of being a scientist.

Starting Out
Scientists answer questions by doing investigations. In an investigation, you may test or try different things. You should do some things before you start an investigation. Listen to your teacher. Follow all the directions your teacher gives you. Read the investigation before you start to work.

As You Work
Be sure to wear safety equipment if your teacher tells you to. Glass can break, and liquids can splash. **Safety goggles** will protect your eyes. A lab apron and gloves will protect your clothes and skin. Always roll up long sleeves. Tie back long hair, too.

Look for safety symbols in an investigation. **Safety symbols** are small pictures that show possible dangers. They also show reasons to be careful. The chart below shows some safety symbols you should know.

Safety Symbols

Danger of electric shock— follow instructions carefully.	Be careful using materials that burn easily.	Be careful using sharp objects.	Wear safety goggles to protect your eyes.	Be careful using poisonous materials. Do not put anything near your mouth.

Duplicating any part of this book is prohibited by law.

Here are some simple rules to follow to stay safe:

- Always use the safety equipment you are told to use. These objects keep you safe while you learn.
- Do not eat or drink while you are doing a science investigation.
- Always be careful with science tools or materials.
- Do not put anything in your mouth. Do not smell anything unless the directions tell you to.
- Always follow instructions in the right order. Changing the order might change the whole investigation.
- Do not touch anything that is hot. Keep your hands and clothes away from flames or hot objects.
- Use care with sharp tools such as scissors. Always point sharp objects away from yourself and others when you use them. Carry sharp objects the right way.
- Do not mix substances together unless your teacher tells you to. Mixing some chemicals can make a new liquid or gas that could be harmful.
- If something spills or breaks, tell your teacher right away. Never clean up broken glass by yourself. Do not try to hide spills or breaks. Your teacher wants you to stay safe and will be glad you reported the problem.

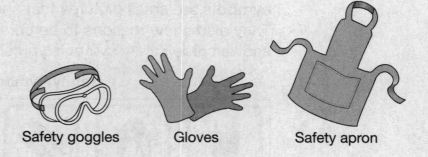

Safety goggles   Gloves   Safety apron

When You Are Finished

After you finish an investigation, clean up. Put away your tools, materials, and safety equipment when your teacher tells you to. Listen to your teacher for directions. Then ask yourself, What did I learn? What should I investigate next?

Lesson 5: **Safety in Science**

Test Tips...

Cross out any answer choices that you know are wrong. Then choose the correct answer from the choices that are left.

DISCUSSION QUESTION

Why do you think you should not eat or drink while you are doing an investigation?

_____

LESSON REVIEW

1. A student spills some liquid from a test tube. What should the student do first?

   A. remove safety goggles
   B. ask another student to help him
   C. tell the teacher
   D. get some paper towels

2. When you see the symbol below, what reason is there for you to be careful?

   A. sharp objects
   B. poison
   C. materials that burn easily
   D. danger of electric shock

3. Which one does NOT protect you when you do an investigation?

   A. goggles
   B. long sleeves
   C. gloves
   D. lab apron

4. You are about to carry out a science investigation. What should you do first?

   A. clean up your desk
   B. set up the equipment
   C. write in your science notebook
   D. listen to directions

Review

1. Earth moves around the sun. This causes the seasons. The sun and Earth form a

   A trend
   B system
   C prediction
   D pattern

2. Earth moves around the sun. This causes the seasons. The seasons are a

   A trend
   B system
   C prediction
   D pattern

3. The graph shows two guinea pigs, A and B, on different diets. Which statement *best* describes the trend you see in the weight of guinea pig A?

   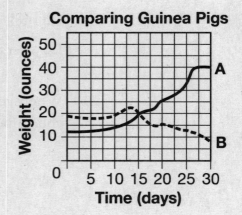

   A It gained weight.
   B It gained weight, then lost weight.
   C It lost weight, then gained weight.
   D It lost weight.

4. Which tool should you use to measure the distance around a pumpkin?

   A graduated cylinder
   B ruler
   C scale
   D measuring tape

5 Which tool is used to measure the volume of a liquid?

  A balance
  B scale
  C graduated cylinder
  D thermometer

6 A student investigates rocks that he finds near his home. Which one could the student use to record what he observes?

A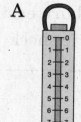

C

B

D

7 Which one is correct?

  A 1 meter = 1000 kilometers

  B 1 centimeter = 100 meters

  C 1 milliliter = $\frac{1}{1000}$ liter

  D 1 kilogram = $\frac{1}{100}$ gram

8 Which choice listed below is a metric unit?

A   gram
B   foot
C   pound
D   gallon

9 When you see the symbol below, what is the reason for you to be careful?

A   sharp object
B   poison
C   materials that burn easily
D   danger of electric shock

10 Is this a question you can test?

> How tall do sunflowers grow?

Identify whether this is a testable question and explain your answer.

_____

_____

_____

11 A student wants to investigate leaf sizes. She measures the length of one leaf. Explain what she should do next.

_____

_____

_____

CHAPTER 2
Experiments

Lesson 6: Design an Investigation

S1.3a; S2.1a, b; S2.3a; IP2

Getting the Idea

Scientists ask questions about the natural world. Then they investigate to try to find answers. Sometimes scientists observe something that changes over time. That is one kind of investigation. Another kind of investigation is a carefully controlled test called an experiment.

Key Words
- observation
- prediction
- investigation
- experiment
- variable
- controlled variable
- independent variable
- dependent variable
- data
- conclusion

Make Observations and Ask Questions

In Lesson 1, you learned about making observations. Recall that an **observation** is information that you gather with your senses. Suppose you observe that a puddle dries up quickly on a sunny summer day. You wonder how water evaporates, or turns from a liquid to a gas. You ask the question, "How does sunlight affect the evaporation of water?"

Make a Prediction

After scientists ask a question, they think of a possible answer. They use what they already know to think of an answer that makes sense. They make a **prediction**—a guess about what is likely to happen or about what is true. You can test a good prediction to see if it is correct. What prediction would you make about how water evaporates? First, think about what you already know about water and sunlight. You might predict that sunlight makes water evaporate at a faster rate.

Lesson 6: **Design an Investigation**

Kinds of Investigations

An **investigation** is a careful study of something to answer a question about it. Scientists use different kinds of investigations. Which kind they choose depends on the question they want to answer.

Sometimes scientists investigate just by observing and collecting data. You might observe where and how a frog lays her eggs. You might observe layers of rock on the side of a cliff. When there is a meteor shower, you might observe one streak across the night sky.

Some investigations are experiments. An **experiment** is a test performed to answer a question in science. Wrapping your sandwich two different ways can be an experiment. But you must control what you test. Then you will know what causes any change.

How would you test your prediction about evaporation? Suppose you decide to fill two bowls with water. You plan to place one bowl in sunlight and the other bowl in shade. The picture below shows the setup for your experiment.

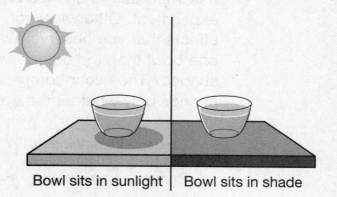

Bowl sits in sunlight | Bowl sits in shade

Part of planning an experiment is deciding what the variables will be. A **variable** is something that can change, or vary, in an experiment. An experiment has different kinds of variables.

Controlled variables are conditions that a scientist keeps the same in all parts of an experiment. In your experiment, you will use two bowls that are the same size and shape. You will put the same amount of water in each bowl. You will let both bowls sit for the same amount of time.

An **independent variable** is a condition that a scientist changes in different parts of an experiment. Whether the bowls get sunlight is the independent variable in your experiment. One bowl of water will get sunlight, and the other bowl will not.

A **dependent variable** is something that changes when a scientist changes the independent variable. The dependent variable *depends* on the independent variable. In your experiment, the dependent variable is the amount of water that evaporates from each bowl.

Suppose that you let the bowls sit for three hours. At the end of that time, you measure the water in each bowl. Then you compare the amounts.

It is important to change only one condition in an experiment. Otherwise, you will not know what causes any effects that you observe. What if you put more water in one bowl than the other? What if the bowls are different shapes? Then your comparisons will not be fair. You must change only whether the bowls get sunlight.

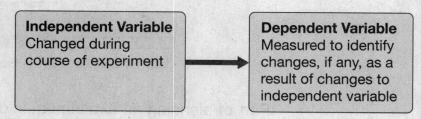

Write a Plan

After you decide what your variables will be, you are ready to write a plan for your experiment. First, list all the materials you will need. Then write the steps you will follow. It is a good idea to share your plan with others. They may have good suggestions for improving your plan.

Record Data and State a Conclusion

When you do an experiment, you must record, or write down, your data. **Data** are pieces of information, often in the form of numbers. Data from an experiment include observations and measurements. The data from your experiment include the amount of water you put in each bowl and the amount of water left after three hours.

You may wish to show your data in a chart. It might look like this one.

Effect of Sunlight on the Evaporation of Water

Location of Bowl	Water at Start	Water after 3 Hours
Sunlight	200 mL	179 mL
Shade	200 mL	195 mL

Next, you must draw or state a conclusion. A **conclusion** is a statement about what you think your data mean. Your data show that more water evaporated from the bowl that was in sunlight. You conclude that sunlight makes water evaporate faster.

Your data support your prediction—the information shows that your prediction was correct. What if the data did not support your prediction? It is fine if your guess was not right. That would not mean that your experiment was a failure. You still learned something. You could do other experiments to learn even more.

DISCUSSION QUESTION

Why is it important to change only one variable in an experiment?

Test Tips...

Cover the answer choices before you read a multiple-choice question. Read the question and try to answer it in your own words. Then read the answer choices. Choose the one that is closest to your answer.

LESSON REVIEW

1. What is the main purpose of an experiment?
   A. to make a prediction
   B. to ask a question
   C. to prove that a prediction is correct
   D. to test a prediction

2. Which sentence is correct?
   A. An experiment should have only one variable.
   B. An experiment begins with drawing conclusions and ends with asking questions.
   C. Doing an experiment includes recording data and drawing conclusions.
   D. You can make up an experiment as you go along.

3. What do scientists call conditions that are kept the same in all parts of an experiment?
   A. controlled variables
   B. independent variables
   C. dependent variables
   D. conclusions

Lesson 7
Collect and Organize Data

M3.1a; S2.3a, b; S3.1a; IS1; CT2, 5

Getting the Idea

Key Words
organize
data
table
bar graph
graph paper
line graph

Suppose that you filled your science journal with information. Now you want to figure out what it means. You recorded measurements on the first page. You recorded some other measurements on page 10. You wrote some more on page 15. You flip back and forth. You need to **organize** your information. You need to arrange it in some kind of order. Then it will be easier to understand.

Use Tables and Graphs

Data, or pieces of information, are often in the form of numbers. Scientists organize numbers in different ways. One way is in a table. A **table** is a chart made up of rows and columns. The table below shows data that one student collected.

Schoolyard Bird Watch

Kind of Bird	Number of Times Seen
Crows	12
Blue jays	8
Chickadees	20
Bluebirds	2
Cardinals	8

The table makes it easier to look at the data. Read across each row. You can see at a glance how often the student saw each kind of bird.

The bar graph below shows the same data in a different way. A **bar graph** is a chart that shows numbers as bars of different lengths.

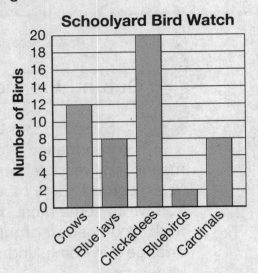

A bar graph makes it easier to compare things. You compare the data by looking at the lengths of the bars. You can see that the student saw chickadees most often.

When you make a graph, you use **graph paper**. Graph paper has lines across and up and down in a grid pattern. The lines help you make the bars the correct length.

Changes over Time

Scientists often want to know how something changes. They collect data over a certain period of time. Then they look for patterns in the data. They may put the data into a table. A table can make it easier to see patterns of change.

A class had a pet guinea pig. It was two months old when the class got it. The students weighed and measured the guinea pig every week for six weeks. They recorded the data. Then they put the data into a table.

How a Guinea Pig Grew

	Week 1	Week 2	Week 3	Week 4	Week 5	Week 6
Length (cm)	20	24	28	30	30	30
Weight (N)	5	7	8	9	10	10

Test Tips...

To answer multiple-choice questions, turn them into true-false questions. Read each choice and ask yourself if it is true. The choice that seems most true is likely the answer.

The table helped the students understand how their guinea pig changed. It got longer for four weeks. It got heavier for five weeks. After week 5, the guinea pig did not get longer or heavier. By then, it was three months old. The students drew a conclusion. They decided that their guinea pig was full-grown when it was about three months old.

A **line graph** is a graph that shows data as a line across a grid. A line graph is a good way to show changes over time. The graph below shows how the weight of the guinea pig changed over six weeks.

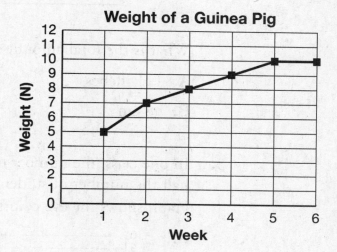

DISCUSSION QUESTION

Why do you think it is a good idea to organize the data from a science investigation?

LESSON REVIEW

1. A bar graph is the BEST choice when you want to

   A. see changes over time.

   B. collect data.

   C. compare things.

   D. measure things.

2. The table below shows the amount of rain that fell in a city in three different months.

Rainfall

Month	Amount of Rain
March	3 cm
April	4 cm
May	4 cm

What is the total amount of rain for the three months?

A. 11 inches

B. 8 cm

C. 4 cm

D. 11 cm

3. In one class, the number of students with blue eyes is half the number of students with brown eyes. Which bar graph shows the eye colors of the students in that class?

A.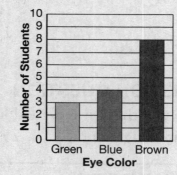

C.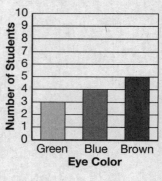

B.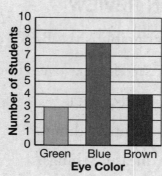

D.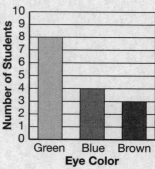

Lesson 8: Figure Out Data

M2.1a; S2.3a; S3.2a; S3.3a; IS3; CT6

Getting the Idea

Key Words
record
analyze
conclusion

Good scientists keep careful records of their investigations. A **record** is information written down and saved for future use. Scientists study their records to figure out what their results mean. Then they draw conclusions that explain the results.

Keep Records

You should always keep careful records during an investigation. If you do not, you may not be able to explain your results. Suppose a scientist investigates a kind of mold that grows on oranges. The scientist asks if light makes the mold grow faster.

He takes the moldy peel off an orange. He puts some of the peel under a bright light. He puts some in a dim corner of his lab and some in a dark closet. At the end of one week, he gathers all three samples. He puts them on his lab table. He looks at each sample and writes his observations in a table like the one below. He compares them to see how light affected the mold.

Effects of Light on Mold Growth

Sample	Light	Observations after 1 Week
1	Bright	No change
2	Dim	Very little mold growth
3	None	Large amount of mold growth

After the scientist records his observations, he analyzes them. When you **analyze** the results of an experiment, you try to figure out what they mean. You look for patterns and trends. You look to see if one variable affected another.

Analyzing the results can lead to a conclusion. Recall that a **conclusion** is a statement about what data from an investigation may mean. In this experiment, the scientist concluded that mold grows best in the dark. Notice that the conclusion is an answer to the question. The scientist asked if light makes mold grow faster. He concluded that light does not make mold grow faster. His data led him to this answer.

A Backyard Experiment

Student A observes what takes place in her backyard. She notices that squirrels steal food from her family's bird feeder. This gives her an idea. She wants to know which kind of bird food squirrels prefer—flaxseed, sunflower seed, or mixed seed.

She sets up three matching bird feeders. She labels the feeders 1, 2, and 3. She fills each feeder with a different kind of food. Then she places the feeders in her yard. She watches them for two hours each afternoon for four days. She records her data in the table shown below.

Number of Squirrels Visiting Bird Feeders

Feeder	Food	Day 1	Day 2	Day 3	Day 4
1	Flaxseeds	6	8	4	9
2	Sunflower seeds	12	12	9	13
3	Mixed seeds	18	22	14	17

The student analyzes her data. When she reads across the rows, the student does not see any pattern. Different numbers of squirrels visited each feeder on different days. But when she reads down the columns, she does find a pattern. Every day, more squirrels came to feeder 2 than to feeder 1. Also, every day, more squirrels visited feeder 3 than the other two feeders.

Lesson 8: **Figure Out Data**

The student draws a conclusion from her data. She concludes that squirrels like sunflower seeds better than flaxseeds, but they like mixed seeds the best.

Repeating the Experiment

Another student, student B, decides to repeat the investigation in his own backyard. He uses similar materials and follows the same steps. He records his results in this data table.

Number of Squirrels Visiting Bird Feeders

Feeder	Food	Day 1	Day 2	Day 3	Day 4
1	Flaxseeds	5	6	8	5
2	Sunflower seeds	15	18	12	14
3	Mixed seeds	10	12	9	11

The student draws a conclusion from his data. He concludes that squirrels prefer sunflower seeds compared to the other birdseeds.

Comparing Results

The students wonder why they got different results. They ask new questions. Was the bird food student B used different from the kind student A used? Were student B's mixed seeds stale? Were student B's sunflower seeds extra tasty? Did the squirrels not like the location of student B's mixed seed feeder?

Should student B change his results to agree with those of student A? No, scientists should always report their results honestly. Scientists must be able to trust each other's results.

Student A and student B should try to figure out why they got different results. They can repeat the investigation or investigate one of the new questions. They can also ask another student to try the same investigation. They still may not find a clear answer to their question, but they will learn more.

Graphs Help You Compare

Graphs can help you analyze data and draw conclusions. A student did an experiment to test different brands of paper towels. She poured one cup of water onto the counter and counted how many paper towels she used to clean up the water. She repeated this procedure using four different brands of paper towels. She recorded her data and used the numbers to make this graph.

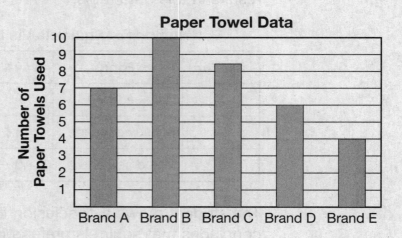

To draw a conclusion, compare the heights of the bars on the graph. Brand B has the highest bar. At first, you might think that this is the best because it has the highest bar. But brand B is the worst because the student needed the most towels to clean up the water when she used brand B. The best paper towel is the one with the shortest bar, brand E. The student used the fewest towels when she used brand E.

DISCUSSION QUESTION

An investigation has these results: Dogs given food A grew faster than dogs given food B. Food A has 10 grams of protein. Food B has 5 grams of protein. Everything else in the two foods is the same. What can you conclude from these results?

Lesson 8: **Figure Out Data**

LESSON REVIEW

1. After you finish an investigation, you state what seems to be true based on your results. What is this statement called?

   A. a conclusion C. a prediction
   B. a record   D. a table

2. A scientist does not keep good records during an investigation. Which of these things is MOST LIKELY to happen?

   A. The scientist will remember his data correctly.
   B. The scientist will save time.
   C. The scientist's prediction will be correct.
   D. The scientist's conclusion will not be correct.

3. A student planted four seedlings in different pots. She gave pots 1 and 2 the same amount of water each day. She did not water pots 3 or 4. She recorded her data in this table.

   Seedling Height

Pot	Water per Day	Day 3	Day 6	Day 9	Day 12
1	50 mL	1 cm	2 cm	$3\frac{1}{2}$ cm	5 cm
2	50 mL	$1\frac{1}{2}$ cm	$2\frac{1}{2}$ cm	4 cm	5 cm
3	None	1 cm	$1\frac{1}{2}$ cm	$1\frac{1}{2}$ cm	Dead
4	None	1 cm	1 cm	Dead	Dead

   Which statement about the data is TRUE?

   A. The seedlings that got no water grew the most.
   B. The seedling in pot 3 grew well without water.
   C. The effects of no water began to show by day 6.
   D. Until day 9, the seedling in pot 3 grew the same amount as the one in pot 2.

Lesson 9: Present Information to Others

S2.3a; S3.3a; S3.4a, b; IS1, 2

Getting the Idea

You know that every science investigation starts with a question. But how does an investigation end? You may think that it ends with a conclusion, or answer. But there is at least one more step. Scientists share what they learn with others. The things they learn are called their **results**. That is how knowledge about science grows.

Communicating

To **communicate** means to share ideas and information. There are many ways to communicate about science. Writing is one way. Scientists write reports about their work. A **report** is a detailed description of an investigation. Other scientists can learn from the reports.

You can write about your own science investigations. You can share your reports with your class and your family. Drawing pictures is another way to share information. Most often, scientists use both words and pictures to tell about their work. You do not have to be an artist to make a good science drawing. You may even use stick figures. The drawing just needs to explain your ideas or what you saw happen.

Key Words
result
communicate
report

Lesson 9: Present Information to Others

Did You Know?

The Brooklyn Children's Museum was the world's first museum made especially for kids. A museum is a great place to share information, gain knowledge, and have fun while you do it.

A good science drawing has labels that tell what the different parts show. One student investigated how a bean plant grows from a seed. She drew the picture below for her report.

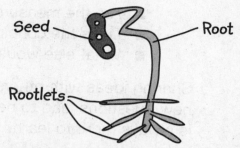

On day 13 the seed looked like this.

In Lesson 7, you learned about using tables and graphs to organize data. Tables and graphs are also good ways to share data. Scientists use tables and graphs to help other people understand their work. You can put tables and graphs in your science reports. A student put the table below in a report about magnets.

What Objects Stick to a Magnet?

Things That Stick	Things That Do Not Stick
Paper clip	Rubber band
Nail	Eraser
Metal spoon	Plastic spoon
Refrigerator door	Wood door

Of course, talking with others is one of the best ways to communicate. Scientists talk about their ideas with other scientists. They present their work at science meetings. They share their ideas on the Internet. You can talk about your ideas with other students in your class. You can give an oral report about a science investigation. Your work can be a source of information for other students.

Asking More Questions

Scientists want to understand each other's work. Be sure to ask questions about someone else's investigation. Here are some more questions you might ask.

- Are the measurements reasonable?
- Does the conclusion make sense?
- What else would I like to know about this?

Sharing ideas with others leads to new questions. And new questions lead to new investigations. In science, there is always more to learn.

DISCUSSION QUESTION

What would happen in a field of study such as medicine if scientists never shared their ideas?

Test Tips...

Be careful when you see the word *best* in a question. More than one answer choice may seem correct. Remember to look for the choice that fits better than any of the others.

LESSON REVIEW

1. Which choice is the BEST definition of the word "communicate"?

   A. to use a telephone
   B. to talk with other people
   C. to write a report
   D. to share ideas and information

2. A drawing for a science report should ALWAYS

   A. explain something clearly.
   B. have bright colors.
   C. be made by a good artist.
   D. look pretty.

3. A student has finished a science investigation. He has written his conclusion in his science notebook. What should he do next?

   A. start doing his ELA assignment
   B. start a brand new investigation
   C. share what he learned from the investigation
   D. keep his ideas to himself

Review

1. Which word means "a guess about what will happen or about what is true"?

   A prediction
   B observation
   C variable
   D conclusion

2. During an experiment, you make and record measurements. What are these numbers called?

   A conclusions
   B data
   C predictions
   D variables

3. You set up an experiment to see how the amount of sunlight affects plant growth. You put one plant in sunlight and the other plant in shade. You give both plants the same amount of water each day. Which condition is a controlled variable?

   A the amount of sunlight
   B the kind of plant
   C the amount of water
   D the amount the plants grew

4 A bar graph of rainfall in April is shown below.

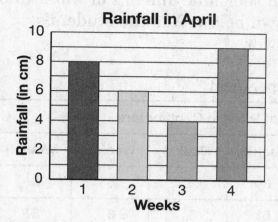

In which week did the most rain fall?

A   week 1
B   week 2
C   week 3
D   week 4

5 A student wants to set up an experiment that is a fair test. What must the student do?

A   change only one condition in different parts of the experiment
B   keep all conditions the same in all parts of the experiment
C   change two conditions in different parts of the experiment
D   figure out a way to make the conditions prove the prediction is true

6 Two students did the same experiment. Student A observed everything carefully but did not write anything down. Student B recorded her observations in a notebook. At the end of the experiment, each student drew a conclusion. Which student is *more likely* to have a correct conclusion?

A   Only student A is likely to have a correct conclusion.
B   Only student B is likely to have a correct conclusion.
C   They both are likely to have a correct conclusion.
D   They both are likely to have an incorrect conclusion.

7 A student predicted that the size of the pot used for planting seeds would not affect the size of the plants. The student used two different sizes of pots. The plants had the same amount of water and light. The student's data table is shown below. Was the student's prediction correct? Explain your answer.

Plant Growth Experiment

Pot	Size	Height of Plant in Centimeters					
		Week 1	Week 2	Week 3	Week 4	Week 5	Week 6
1	small	2	$3\frac{1}{2}$	5	7	8	$8\frac{1}{2}$
2	small	2	$3\frac{1}{2}$	5	$6\frac{1}{2}$	8	$8\frac{1}{2}$
3	large	2	4	5	$7\frac{1}{2}$	10	13
4	large	2	$3\frac{1}{2}$	5	$7\frac{1}{2}$	$10\frac{1}{2}$	$13\frac{1}{2}$

_____

_____

_____

_____

8 In the experiment above, what was the independent variable?

_____

What were **two** of the controlled variables?

(1) _____

(2) _____

CHAPTER 3
Engineering Design

Lesson 10: Technology and Engineering

T1.1a, b; IS3

Getting the Idea

Key Words
technology
engineer
invention
manufacture

Do you ride a bicycle? Do you watch TV? Do you use a computer? Every time you do any of these things, you are using technology. **Technology** is the use of science to solve problems and make people's lives easier. An **engineer** is a person who works on new technology.

Using Technology

Maybe you have heard people talk about things that are "high tech." This usually refers to things that are new and modern. But technology does not have to be complicated. A computer is technology. But a ballpoint pen or a can opener is technology, too.

Hundreds of thousands of years ago, early humans invented simple stone tools. To invent something means to think of it or make it for the very first time. The stone tools, such as knives and hammers, helped make work easier. These tools were some of the first kinds of technology.

Science and technology develop together. Understanding more about science helps people come up with new kinds of technology. A new technology is called an invention. An **invention** is something useful that is made for the first time.

Transportation

Long ago, travel was slow. People used their feet, horses and wagons, or boats pushed by the wind. The slow speed made it hard for people to travel. A trip might take weeks or months. Then, in the 1700s, people used science to invent the steam engine. Steam engines let people travel faster—by train or steamship.

Lesson 10: Technology and Engineering

Did You Know?

Thomas Adams invented chewing gum in New York City in 1870.

Today, there are even faster ways to get from one place to another. Cars and trucks can move people and goods hundreds of miles in just one day. Jet planes travel thousands of miles in a few hours.

Communication

Technology has made communication much faster and easier, too. It used to take days or weeks to send a message to someone who lived far away. People wrote letters by hand, and the letters traveled slowly. In the 1800s, the telegraph and the telephone changed that. These two inventions used electrical energy to send messages very quickly.

Today, we use radios, TVs, cell phones, and computers to stay connected. Messages travel in seconds instead of days or weeks. Satellites that circle Earth help beam messages around the world. The picture below shows a satellite dish. Satellite dishes collect information bounced off satellites in space.

Manufacturing

Technology helps people **manufacture** things, or make products. In the early 1900s, Henry Ford invented a better way to make cars. A moving belt carried the cars along an assembly line. Workers along the line put the cars together part by part. Today, factories use moving assembly lines to make many products. Some of the workers today are robots!

Health Care
Technology also helps keep us healthy. Doctors today can see inside a patient's body without doing surgery. They use tools such as X-ray machines and MRI scanners to see broken bones and other problems.

Farming
Machines help farmers grow and harvest food. Long ago, farmers used hand plows or plows pulled by horses. Today, farmers use machines that can plow fields much faster and more easily. Machines also help farmers plant seeds and pick crops. Then technology such as refrigerator trucks helps get fresh food from farms to supermarkets.

Gathering Information
Technology helps scientists study places that people cannot visit. Scientists use robots to explore the deepest parts of the ocean. Scientists also use robots to explore the insides of caves and volcanoes.

Scientists at NASA have sent spaceships called probes to all the planets in our solar system. The probes do not carry humans. But cameras and other tools send information back to Earth.

Technology Solves Problems
Some problems are big. How do scientists deliver supplies to the International Space Station? The space shuttle solves that problem. Other problems are small. How can you keep a message from getting lost? Using a sticky note solves that problem. When there is a problem, somebody will use technology to invent a solution. The diagram below shows the steps in using technology to solve a problem. You will learn more about these steps in Lessons 11 through 13.

Steps in Using Technology to Solve a Problem

Lesson 10: Technology and Engineering

Test Tips...

When you look at answer choices, look for two that are opposites. One of them has to be wrong. Read the question again to see which answer you can cross out.

DISCUSSION QUESTION
Describe three ways you have used technology today. What is the technology? How does it solve a problem or make life easier?

LESSON REVIEW

1. Which sentence is TRUE?
   A. Only scientists use technology.
   B. Everyone uses technology.
   C. People long ago did not use technology.
   D. Only new inventions are technology.

2. Which technology helps people communicate?
   A. steam engine
   B. computer
   C. assembly line
   D. space probe

3. Technology does NOT help people
   A. send messages.
   B. grow crops.
   C. travel.
   D. think.

Lesson 11 Study a Design

T1.1a, b; T1.2a, b, c; IP1

Getting the Idea

Key Word
design

Technology solves problems. Before you can invent a new technology, you must understand the problem. You also must understand the technology that is being used. To do that, you study how the technology is designed.

Find a Problem

Your class is helping start a community garden. The soil has to be turned over and raked to take out rocks and weeds. You see tools everywhere. There are shovels, rakes, hoes, and some things you have never seen.

Lesson 11: Study a Design

Did You Know?

Thomas Edison, a great American inventor, said, "There's a way to do it better—find it." Not all inventions are completely new ideas. Many are improvements on things that already exist.

At the end of the day, people take their tools home. There's so much to carry! The tools seem very heavy. This gets you thinking. How could tools be changed so that they would not be too heavy to carry?

New technology begins when someone has a problem and wants to solve it. You found a problem—the tools are too heavy. What is the reason? Now you need to figure out why they are so heavy.

Research the Design

Before you can design a new tool, you need to learn more about the tools people use for gardening. To **design** something means to make a plan for solving a problem. You decide to do some research about garden tools.

Where can you get information about garden tools? You look at garden catalogs. You look at gardening sites on the Internet. You go to the library to look for gardening books. You ask your parents to take you to a garden center. Now you have seen even more tools. But you don't know how to make them lighter.

Maybe your grandmother can help you. You remember that she loves to work in her garden. Your grandmother shows you her garden tools. She has a rake, a shovel, and a hoe. She has a tool that makes the edge of a lawn neat. She also has some small hand tools. Her tools look a lot like the ones you saw at the community garden.

It may seem surprising, but studying old technology can help you come up with new ideas. You ask your grandmother what kind of garden tools people used when she was your age. Your grandmother shows you some other tools. They belonged to her grandfather. She does not use them any more. People used tools like these a long time ago.

You compare the old tools to the new tools. The tools are the same kinds of tools you saw at the community garden. There is a rake, a shovel, and a hoe. The old tools are heavier. You see that the metal parts are different. The tools were made of iron. The new ones are made of steel. Steel can be made thinner than iron, so less metal is needed to make a tool. That makes the tool lighter.

The table below compares old and new gardening tools.

Comparing Old and New Designs

Types of Tools	What Kinds Are There?	What Are They Made Of?
Old tools	Rake, hoe, shovel	Iron with wooden handles
New tools	Rake, hoe, shovel	Steel with wooden handles

DISCUSSION QUESTION

Who else could give you information to help you understand the design of garden tools? What would you ask the person? Explain why this person would be helpful.

Lesson 11: Study a Design

LESSON REVIEW

1. What is the first step in inventing new technology?
   A. Find a problem.
   B. Research a design.
   C. Talk to people who know about the design.
   D. Look for a way to solve a problem.

2. What is the problem with the garden tools used at the community garden?
   A. They are too old.
   B. They are broken.
   C. They are heavy.
   D. They are too expensive.

3. A plan to make something new or to solve a problem is called
   A. a problem.
   B. old technology.
   C. your invention.
   D. a design.

Lesson 12 Figure Out New Ideas

T1.1c; T1.3a, b, c; T1.4a, b; IS2; CT2; IP1, 2

Getting the Idea

You figured out one of the problems with garden tools the way they are right now. The tools are too heavy. You looked at the designs of old and new tools. You compared the designs. Now you are ready to try to solve this problem.

Key Words
solution
consumer
revision

Suggest Solutions

Once you understand the problem and the design of the tool, you should start to look for solutions. A **solution** is an answer to a problem. Don't expect to come up with a perfect solution right away. The important thing is to make suggestions and see if they will work. Even if an idea does not work, you may learn something from it or get another idea. If one idea does not work, try another one.

You learned that tools were made lighter by changing from iron to steel. This makes you wonder if another metal could make the tools even lighter. It is time to do more research. You look at things made of metal. You know that an aluminum soda can is lighter than a steel soup can. Maybe aluminum would be good for making garden tools. But is aluminum strong enough?

You try comparing the cans to see if the metal bends. If you squeeze an aluminum soda can, it bends. You can't bend a steel soup can. So maybe aluminum is not a good choice. A rake that bends is no help in a garden!

Aluminum can

Steel can

78 Duplicating any part of this book is prohibited by law.

Lesson 12: Figure Out New Ideas

Did You Know?

Thomas Edison said "Genius is one percent inspiration and ninety-nine percent perspiration." He meant that coming up with a good idea is only a small part of inventing. Most of the effort is all the hard work you put into figuring out how to make it work!

Every design has some limits. You know a rake has to be strong, so you cannot use a metal that bends easily. But you do not want it to be so heavy that you cannot even lift the tools. You need to use a metal that is strong and lightweight.

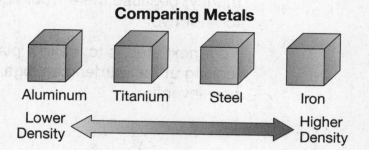

Comparing Metals

Aluminum Titanium Steel Iron

Lower Density ←——————→ Higher Density

Your friend has a light, strong bike made of a metal called titanium. Maybe titanium would be good for making garden tools. You do research and find out how expensive titanium is. It seems a titanium rake would cost a lot of money. It is one thing to spend a lot of money on something like a bike, but you figure nobody wants to spend so much money on garden tools.

You may want to sell your invention someday. If so, you must think about the **consumer**, the person who might buy it. Cost is an important limit. Nobody buys a new invention if it is too expensive.

Make Revisions

If your first ideas do not work, you need to change your solution. Making **revisions**, or changes, in a solution is part of inventing a new technology.

You decide that changing the metal is not the solution to the problem. So then you go back to your original question. How might you change tools so that they are not too heavy to carry? Maybe the problem is not how heavy each tool is. Maybe it is that there are so many tools to carry. Is there one kind of tool that can do the job of two tools? It is time for more research.

You go back to your garden catalogs. You notice that the handles of rakes and hoes look the same. What if you could put the rake on one end of the handle and the hoe on the other end? That would get rid of some of the weight of the set of garden tools. It would also save money, because there would be only one handle to pay for, not two.

Your next step is to make a plan for your tool. After looking at the garden catalogs, you draw a picture of your new invention.

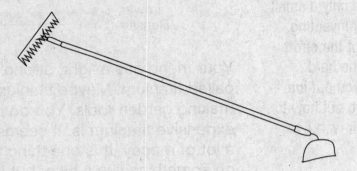

You show your sketch to your parents, and they are impressed. They agree to help you do the next step, which is to build a sample. For some inventions, a small model is enough. For others, a full-size sample is better. Since you already have a rake and a hoe, it makes sense to build a full-size sample.

Before you can build a model or a sample, you need to know how large each part is. You must know what sort of materials you will use. You should also know how the pieces will go together. After you do that, you go on to an exciting step—testing your design. You will learn more about that in the next lesson.

Lesson 12: **Figure Out New Ideas**

DISCUSSION QUESTION
If an idea doesn't work, why isn't it a failure?

LESSON REVIEW

1. What is a consumer?
   A. someone who buys something
   B. someone who invents something
   C. someone who finds a problem
   D. someone who makes revisions

2. Which one is NOT a limit that affects the design of a product?
   A. cost
   B. safety
   C. how well it works
   D. how well the designer can draw

3. Which one is NOT a way to let people know what your new invention will be like?
   A. make a model
   B. make a sample
   C. draw a sketch
   D. worry about the problem

Lesson 13 Test Your Ideas

T1.5a, b, c; CT2, 6; IP1

Getting the Idea

Key Words
prototype
analyze
evaluate

Things are going well with your ideas about garden tools. You came up with what you think is a very good idea. You sketched your design. Then, with the help of your family, you built a sample of your invention. The first sample of an invention is called a **prototype**. Inventors test prototypes to see if their inventions work the right way.

Testing

Now you have your new garden tool. It is time to test it and see how it works. How are you going to test your tool? What will you look for when you test it? You need to see if the tool will do its job of moving soil. You need to see if it will be strong enough. You need to see if it will be safe to use.

You take the tool to the community garden. You ask a friend to take pictures while you try out your tool. First, you try using the rake end. It works well and you are pleased. Then you turn it around to use the hoe end. You hold the rake end up. But the rake is too wide. It is hard to keep the tool straight when you pull back on the hoe. It keeps turning in your hand when you do not want it to turn.

Lesson 13: Test Your Ideas

After you test a prototype, the results must be analyzed. When you **analyze** test results, you study them carefully to understand them. You figure out what happened and what it means. You look at the pictures your friend took. You talk to the people who watched you use the tool.

You must **evaluate**, or judge, your tool based on the test. Is the tool good enough to use? Your test showed that your tool was strong enough to pull through soil without breaking. Your test showed the handle was a good length. Your test also showed that the shape of the rake made the tool hard to use. The tool turned in your hands. It was hard to control. You decide you can do better than that.

Making Revisions

If one idea does not work, try another idea. You look at your prototype to see if there is a way to change it. You talk with your friend at the garden center and discuss ways you could fix the problem. One possibility is to make the rake narrower. Another idea is to put the rake and hoe on the same end, back to back.

You talk some more. After a while, another idea comes to you. You might make different ends that you could take off and switch around. One handle could be for lots of tools. You cannot wait to get home and sketch these new ideas!

DISCUSSION QUESTION

How is testing a prototype like doing an experiment?

LESSON REVIEW

1. What is a prototype?

    A. a revised design

    B. the first sample of a design

    C. the first test of a design

    D. the finished design

2. Which question was NOT used to evaluate the garden tool?

    A. Is it strong enough?

    B. Does it move soil?

    C. Is it safe to use?

    D. Does it look shiny?

3. What should you do if your design does not work well?

    A. give up

    B. find something else to design

    C. look for a way to make your design better

    D. use the design you have

Chapter 3 Review

1. A company plans to make a new product. It has these design limits:

   - It must be strong.
   - It must hold its shape when it is heated.
   - It must be reflective (shiny).

   What is the *best* material for this product?

   A plastic
   B wood
   C paper
   D metal

2. The diagram below shows a crash-test dummy. Scientists use the dummy to test designs for cars. What can the scientists learn from these tests?

   A whether the cars will be safe
   B whether the cars will go fast
   C whether people will buy the cars
   D whether the cars will be comfortable

3 Which one is the *best* definition of the word *design*?

A to make a drawing of something

B to think of a problem to solve

C to make a plan for something

D to test something

4 Suppose you like to do your homework near an open window. Sometimes your papers blow around. You want to design something that solves the problem. Which of these steps should you do *last*?

A describe your solution

B think of some possible solutions

C tell what the problem is

D plan your solution

5 Which technology helped improve manufacturing?

A television

B cell phone

C assembly line

D space shuttle

6 A designer tests a sample of a new product. The sample does not work as planned. What should the designer do?

A throw away the design for the product

B order a factory to make the product anyway

C find a different problem to solve

D make changes in the design

7 A designer has made a design for a new product. What should the designer do next?

    A find a company to make the product

    B test the design

    C communicate the design

    D revise the design

8 What is the first step in using technology to solve a problem?

    A thinking of possible solutions

    B designing a solution

    C testing a solution

    D telling what the problem is

9 Before the telegraph and telephone were invented, people communicated by letter. What is the main problem the telegraph and telephone solved?

    A A letter took a long time to arrive.

    B Stamps for mailing letters were expensive.

    C People didn't like to read letters.

    D A letter wasted paper.

10 You want to design a new backpack that has more pockets to keep pencils, notebooks, and other things organized. What should you do before you start designing?

    A look at the designs of notebooks in stores
    B look at the designs of other backpacks in stores
    C choose a color for your notebooks
    D put away your pencils and notebooks

11 The pump you use to pump up your bicycle tires is hard to use. You want to design a better tire pump. Predict **three** sources of information you could use to research tire pump designs.

(1) _____

(2) _____

(3) _____

CHAPTER 4
Earth's Movement in Space

Lesson 14 Day and Night

PS1.1a, c

Getting the Idea

You get up in the morning, and it is daylight. Hours later, about the time you are having dinner, it is dark. This pattern repeats day after day. Earth's rotation, or spinning, causes the pattern of day and night.

Key Words
axis
rotate
sunrise
horizon
sunset

Why There Is Day and Night

Look at the diagram below. An imaginary line, called Earth's **axis**, runs through Earth from the North Pole to the South Pole. Earth **rotates**, or spins, on its axis. This motion is like that of a spinning top. Earth rotates once every 24 hours. That 24-hour time period is one day.

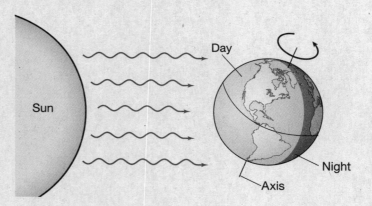

At any time, half of Earth faces the sun. The other half faces away from the sun. In the daytime, the half of Earth that you are on faces the sun. At night, the half of Earth that you are on faces away from the sun.

At the start of each day, the part of Earth you are on spins to face the sun. This time is **sunrise**. The sun appears on the eastern horizon. The **horizon** is the line where the sky and Earth's surface seem to meet. During the day, the sun seems to move across the sky from east to west.

Lesson 14: Day and Night

Did You Know?

On just one day in March and one other day in September, there are exactly 12 hours of daylight and 12 hours of night. Those are the only two days of the year where the numbers match exactly.

In the afternoon, as Earth continues to spin, the sun seems to move lower and lower in the sky. The sun finally reaches the horizon in the west. This time of day is **sunset**, the end of daylight. The sun seems to set in the west as the part of Earth you are on spins to face away from the sun.

The drawing below shows where the sun appears at different times of day. It looks as if the sun rises in the east, moves across the sky, and sets in the west. However, the sun does not really rise in the east. It does not really move across the sky. The sun only seems to move in that way because Earth rotates. If you stand in the middle of a room and slowly spin around, you will see the walls of the room move past you. You know that you are moving and the room is not. But it looks as if the room is moving around you.

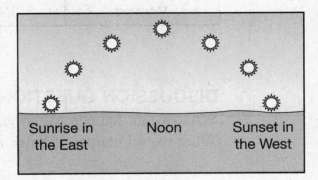

After the sun sets, the sky is dark and you can see stars in the sky. During the night, most stars look as if they move across the sky from the east to the west. The stars do not really move the way they seem to move. Like the sun, they seem to move in that way because Earth is turning on its axis. As Earth continues to spin, the sun will rise again the next morning. The cycle will repeat.

Shadows

When light strikes an object, the object blocks the light. A dark area, called a shadow, appears on the side of the object that faces away from the light. As the sun shines from different directions at different times, shadows change in length and position. You can use the length and position of shadows to tell the time of day.

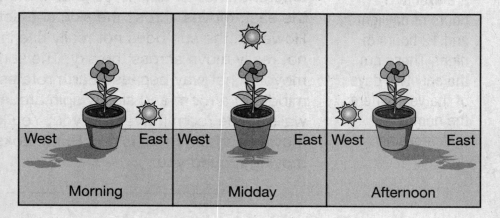

DISCUSSION QUESTION

What would happen to day and night if Earth rotated faster? What would happen if Earth rotated more slowly?

LESSON REVIEW

1. Around what does Earth rotate?

    A. the sun

    B. its axis

    C. the equator

    D. its horizon

2. How long does Earth take to rotate once?

    A. 12 hours

    B. 20 hours

    C. 24 hours

    D. 36 hours

3. As Earth spins, the part facing the sun has

    A. spring.

    B. darkness.

    C. daylight.

    D. summer.

4. Where does the sun set?

    A. in the north

    B. in the east

    C. in the south

    D. in the west

Lesson 15 Seasons

PS1.1a

Getting the Idea

As the seasons change, you notice changes in the weather. Summer in New York is often very warm. Winter is cold. You might think that Earth is closer to the sun in summer than in winter. But this is not true. Earth's tilt on its axis and its movement around the sun cause the seasons.

Key Words
equator
hemisphere
revolve
orbit

Causes of Four Seasons

Why is it warmer in summer than in winter? Recall that Earth rotates on its axis, like a top. When a top spins, the axis is straight up and down. But Earth is not a top. Earth's axis is not straight up and down. Instead, it tilts. This means part of Earth leans toward the sun while another part leans away.

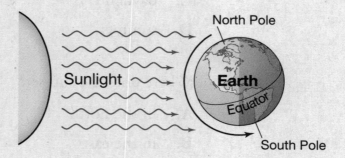

Earth can be divided in half by an imaginary line. The **equator** is an imaginary line around the middle of Earth. Each half of Earth is a **hemisphere**. The half of Earth above the equator is the Northern Hemisphere. The half of Earth below the equator is the Southern Hemisphere.

When the Northern Hemisphere tilts toward the sun, it is summer in the Northern Hemisphere. The sun's rays heat the Northern Hemisphere more than the Southern Hemisphere.

Lesson 15: Seasons

Earth moves in two ways. You know that Earth rotates, or spins on its axis. This causes night and day. Earth also **revolves**, or moves in a path around the sun. This path around the sun is its **orbit**. Earth takes one year to go all the way around the sun one time.

Look at the diagram below. You can see that Earth's axis tilts. As Earth revolves, different parts of Earth tilt toward the sun. Notice the tilt when it is summer in the Northern Hemisphere. The northern half of the planet tilts toward the sun. Now notice Earth's tilt when it is winter in the northern half of Earth. The Northern Hemisphere tilts away from the sun.

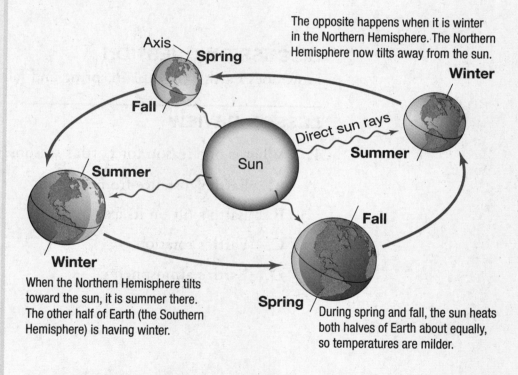

The opposite happens when it is winter in the Northern Hemisphere. The Northern Hemisphere now tilts away from the sun.

When the Northern Hemisphere tilts toward the sun, it is summer there. The other half of Earth (the Southern Hemisphere) is having winter.

During spring and fall, the sun heats both halves of Earth about equally, so temperatures are milder.

Look at the seasons in the Southern Hemisphere. When it is summer in the Northern Hemisphere, it is winter in the Southern Hemisphere. When it is winter in the Northern Hemisphere, it is summer in the Southern Hemisphere. In places near the equator, the weather does not change much from one season to the next.

The other two seasons are spring and fall. During these seasons, neither hemisphere tilts toward the sun. The sun heats the two hemispheres evenly. Temperatures are milder.

The length of daylight each day also changes along with the seasons. This change affects the temperature. In summer, there are more hours of daylight. So when the sun's rays are more direct, the sun also heats Earth's surface longer each day. In winter, conditions are just the opposite. The sun's rays are less direct, and there are fewer hours of daylight. Both conditions together cause the lower temperatures of winter.

DISCUSSION QUESTION

How does Earth's tilt explain spring and fall here in New York?

LESSON REVIEW

1. What is one reason for Earth's seasons?

    A. Earth's distance from the sun

    B. Earth's tilt on its axis

    C. Earth's rotation

    D. Earth's atmosphere

Lesson 15: **Seasons**

Test Tips...

Some questions are easier to answer if you can see a drawing that helps explain what is being asked. Draw a diagram on scrap paper if you need it. The diagram can help you understand the question and figure out the answer.

2. When the northern half of Earth tilts toward the sun, that part of Earth has

   A. fall.
   B. spring.
   C. winter.
   D. summer.

3. In winter in New York, the sun's rays hit the Northern Hemisphere

   A. only at noon.
   B. less directly than in summer.
   C. more directly than in summer.
   D. 24 hours each day.

4. In winter in New York, the Northern Hemisphere is

   A. tilted away from the sun.
   B. the same distance from the sun as it is in summer.
   C. closer to the sun than in summer.
   D. always in darkness.

Lesson

16 Our View of the Moon

Getting the Idea

Sometimes the moon looks full and round. At other times, it is only a small sliver in the sky. In this lesson, you will learn why the shape of the moon seems to change.

Key Words
phase
reflect
waxing
waning

Moon Phases

Phases are the different shapes of the moon that we see from Earth. The moon itself is always a round ball. That shape does not change. What does change is how much of the lit part of the moon we can see from Earth.

The moon is the brightest object in the night sky. But the moon does not give off its own light. The moon reflects the light of the sun. When an object **reflects** light, the light bounces off the object's surface. The sun always shines on one-half of the moon. The shapes that we see are parts of that lit half. The picture below shows the four main phases of the moon.

Four Main Phases of the Moon

New moon   First quarter   Full moon   Last quarter

A new moon happens when the bright side of the moon faces away from Earth. All we see is the dark side. Most often, a new moon looks like no moon at all. A full moon is the opposite of a new moon. It looks like a bright, round circle. During a full moon, we see the entire bright side. During the first quarter, we see half of the lit side of the moon. The moon looks like a half-circle. During the last quarter, we see the opposite half of the moon's lit side. Again, the moon looks like a half-circle.

The moon also has crescent and gibbous phases. During a crescent phase, we see just a sliver of the bright side of the moon. During a gibbous phase, we see about three-fourths of the bright side.

Why Do We See Phases?

Look at the diagram below. You can see that sunlight always lights half the moon's surface. But the moon does not stay in one place. It moves in an orbit around Earth. As the moon moves, we see different parts of the half that is lit. We see different phases of the moon.

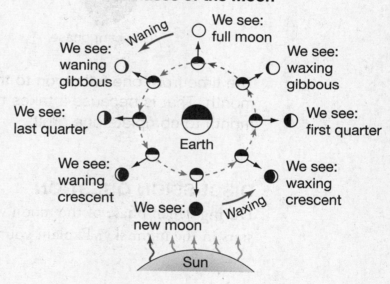

Phases of the Moon

In the diagram, the inside circle shows the moon revolving around Earth. The outside circle shows the phases that we see from Earth in different parts of the moon's orbit. Remember that Earth is rotating, so your part of Earth faces the moon once in each 24 hours. No matter where you are on Earth, the moon's phase will look the same.

Order of Moon Phases

From new moon to full moon, the lit part of the moon that we see gets larger. We say that the moon's phases are **waxing**, or getting bigger. From full moon back to the next new moon, the lit part of the moon that we see gets smaller. We say that the moon's phases are **waning**.

As the moon waxes, the phases go from new moon, to waxing crescent, to first quarter, waxing gibbous, and then full moon. As the moon wanes, you first see a waning gibbous phase. After that, the phases are last quarter, waning crescent, and then new moon again.

Crescent phase   Gibbous phase

The time from one full moon to the next is about one month. That is because it takes the moon about one month to complete one orbit.

DISCUSSION QUESTION

During which phase of the moon would it be easiest to see stars in the night sky? Explain your answer.

Lesson 16: Our View of the Moon

Test Tips...

If two choices seem correct, read the question again and compare those choices for differences.

LESSON REVIEW

1. During which phase does the moon look like a bright, round circle in the sky?

   A. first quarter

   B. new moon

   C. last quarter

   D. full moon

2. What is the main source of the moon's light?

   A. Earth

   B. the sun

   C. the moon

   D. the stars in the night sky

3. Which phase of the moon is shown in the picture?

   A. waxing crescent

   B. full moon

   C. last quarter

   D. waning gibbous

4. What phase comes right before full moon?

   A. new moon

   B. last quarter

   C. waxing gibbous

   D. waning gibbous

Lesson 17 Measuring Time

PS1.1b

Getting the Idea

Key Words
year
day
week
month
calendar
hour
minute
second

How many hours did you sleep last night? How many days are there until your birthday? How many seconds did it take a runner to run 50 meters? How are these questions alike? These are all about different periods of time.

Days, Weeks, Months, and Years

Two measurements of time are based on ways that Earth moves. One **year** is the time it takes Earth to complete one full orbit around the sun. One **day** is the time it takes Earth to rotate on its axis once. There are $365\frac{1}{4}$ days in one year.

Other time measurements are not based directly on Earth's motion. These are units people invented to organize time. One **week** is seven days. Another unit of time that people made is the month. One year is divided into 12 **months**. Months are not all the same length. The shortest month, February, has 28 or 29 days. All the other months have 30 or 31 days.

January

	Mon	Tue	Wed	Thu	Fri	Sat	Sun
1		1	2	3	4	5	6
2	7	8	9	10	11	12	13
3	14	15	16	17	18	19	20
4	21	22	23	24	25	26	27
5	28	29	30	31			

● = Full moon ● = New moon

102 Duplicating any part of this book is prohibited by law.

Lesson 17: Measuring Time

People use a chart called a **calendar** to keep track of days, weeks, and months. Each row is a week. Each square shows one day. You can mark important dates on a calendar so you will not forget them. Most calendars have one page for each month. Calendars often label holidays.

Parts of a Day

People use units of time to show parts of a day. One day is divided into 24 **hours**. Each hour is divided into 60 **minutes**. Each minute is divided into 60 **seconds**. People use a clock or a watch to keep track of hours and minutes. Some clocks and watches show seconds, too. Clocks can show time on a dial or as a set of numbers.

Clock   Digital clock   Stopwatch

Choosing Time Units

Look at the questions below.

- How many hours did you sleep last night?
- How many months are left until your birthday?
- How many seconds did it take to finish the race?

You can figure out the answers to questions like these. But what if you have the same questions asked in a different way? Look at the questions below. How do the two sets of questions compare?

- How many weeks did you sleep last night?
- How many seconds are left until your birthday?
- How many years did it take to finish the race?

Did You Know?

You share your birthday with even more people than you realize. Around ten million people all around the world were born on the same day as you.

The first set of questions on page 103 makes sense. The questions are easy to answer. The second set of questions does not make sense. This is because the units are not good choices. Your birthday might be exactly 10 days away. That is an easy number. If you change 10 days to seconds, then it is 864,000 seconds until your birthday. That is a much harder number to figure out!

When you choose a time unit, you want a unit that fits the amount of time. That is why you give your age in years, but the length of a song in minutes. Some scientists study things that happen so fast that even a second is not a small enough unit. They use units that are parts of seconds. For example, a millisecond is one-thousandth of a second.

DISCUSSION QUESTION

What would be the best time unit for telling how long you take to eat lunch? What would be the best time unit for telling how long the school day is?

Lesson 17: **Measuring Time**

LESSON REVIEW

1. What unit of time is highlighted below?

Mon	Tue	Wed	Thu	Fri	Sat	Sun
	1	2	3	4	5	6
7	8	9	10	11	12	13
14	15	16	17	18	19	20
21	22	23	24	25	26	27
28	29	30	31			

   A. 24 hours

   B. 1 day

   C. 1 week

   D. 14 days

2. Which one is the shortest amount of time?

   A. one hour

   B. one second

   C. one day

   D. one month

3. How long is one year?

   A. 12 weeks

   B. 100 days

   C. $365\frac{1}{4}$ days

   D. 7 months

Chapter 4 Review

1. Earth's tilt on its axis causes

   A its rotation
   B seasons
   C its revolution
   D day and night

2. When it is winter in the Northern Hemisphere,

   A Earth is farther from the sun
   B the Southern Hemisphere tilts toward the sun
   C the sun's rays are more direct than in summer
   D the Northern Hemisphere tilts toward the sun

3. Which statement is correct?

   A The moon reflects light from all the stars.
   B The moon gives off its own light.
   C The moon reflects light from the sun.
   D No one knows the source of the moon's light.

4. How much of the moon's surface is always lit at one time?

   A One-fourth
   B One-half
   C Three-fourths
   D All

5 The picture below shows the moon and stars in a night sky.

Which phase of the moon would come about two weeks *after* the one you see in the picture?

A last quarter
B first quarter
C waning gibbous
D new moon

6 When the lighted part of the moon seems to get larger, the moon is said to be

A waning
B new
C waxing
D third quarter

7 What causes the position of the sun in the sky and shadows to change throughout the day?

A Earth revolving around the sun
B Earth rotating on its axis
C the sun shining on the moon
D the moon going through its phases

8 Earth revolves around the sun in one

A day
B week
C month
D year

9 Which one is the longest amount of time?

   A   one month
   B   one week
   C   one day
   D   one hour

10 The diagram below shows Earth in four positions as it revolves around the sun.

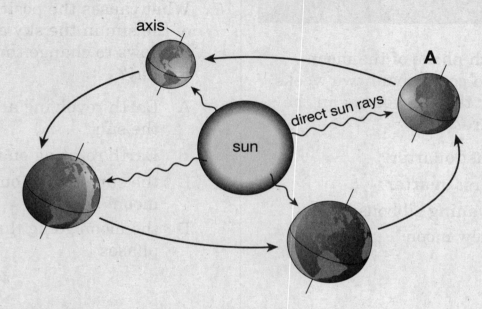

When Earth is at the position labeled *A*, what season is it in the Northern Hemisphere? What season is it in the Southern Hemisphere? Describe how you know.

_____

_____

_____

CHAPTER 5: Planet Earth

Lesson 18 Weather

S2.3a; PS2.1a, b; PS3.1e

Getting the Idea

Key Words
weather
temperature
precipitation
thermometer
wind vane
anemometer
rain gauge

When you get up in the morning, you look out the window. You want to know about the weather that day. **Weather** is the condition of the outside air at a certain time and place. Sometimes the weather is cold. At other times, it is warm. There could be rain or snow. There are many types of weather, and it is possible to figure out when certain kinds of weather will probably happen.

Weather Patterns

The weather in New York can change from day to day. It also changes with the seasons. Each season has a certain type of weather. Certain temperatures are a part of the weather for each season. **Temperature** is a measure of how warm something is. The temperature of the air is coldest in winter. In New York, January is usually the coldest month of the year. The coldest it usually gets in Albany, New York, is 13 degrees Fahrenheit (°F).

The temperature is warmest in summer. July is usually the warmest month in New York, with temperatures such as 83°F in Albany. In spring and fall, the weather is mild. It is not as hot as in summer or as cold as in winter.

110 Duplicating any part of this book is prohibited by law.

Lesson 18: **Weather**

Did You Know?

There are names for different types of clouds. Light, wispy clouds high in the air are cirrus clouds. Fluffy clouds you might see on a nice day are cumulus clouds. Big, thick blankets of clouds are stratus clouds.

The temperature can tell you how it will feel when you go outside. It can also tell you the type of precipitation that may fall. **Precipitation** is water that falls from clouds to Earth's surface. It can be rain, a liquid. It can also be a solid—snow, sleet, or hail. As the temperature gets colder, the chance of solid precipitation is greater. About 36 inches of precipitation falls in Albany, New York, each year. In the coldest parts of winter, much of it falls as snow and sleet.

The table below describes the four main categories of precipitation.

Different Kinds of Precipitation

Precipitation	Description
Rain	Drops of liquid water that fall from clouds. Tiny droplets of water come together to form raindrops large enough to fall to the ground.
Snow	Ice crystals that form in cold clouds. Snow can take the shape of six-sided flakes or beads of ice.
Sleet	Frozen or partly frozen rain. Sleet forms when rain falls through a layer of the atmosphere that is above freezing temperatures, and then through a freezing layer closer to the ground.
Hail	Balls of ice that form in thunderclouds

Figuring Out Weather

You can find many weather patterns just by looking outside and thinking about the seasons. But you probably learn about the weather most often from weather forecasters on TV or the radio. Forecasters tell you what the temperature will probably be. They tell you how fast the wind is blowing. They know how much rain has fallen in a day—and if rain may fall tomorrow. How do they know so much about the weather? Certain tools give weather scientists information about the weather. These tools help them measure the properties of the weather, such as temperature and precipitation.

Measuring the Weather

Scientists use a **thermometer** to measure temperature. There are several kinds of thermometers. The most common has a liquid inside a thin glass tube. The liquid expands, or takes up more space, as it gets warmer. It contracts, or takes up less space, as it gets cooler. As the liquid expands and contracts, it goes up and down in the tube. The changing level of the liquid shows the temperature.

Weather tools describe two properties of wind. They show its direction and its speed. A **wind vane** shows which way the wind blows. You might see a wind vane on the roof of a house or a barn. A wind vane spins freely in the wind. The front points in the direction the wind comes from. If the wind vane points east, that means the wind comes from the east.

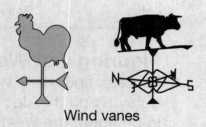

Wind vanes

Lesson 18: Weather

An **anemometer** tells how fast the wind blows. This tool has several arms that stick out. Each arm has a cup at its end. Wind blows into the cups and spins the anemometer. The stronger the wind, the faster the arms spin.

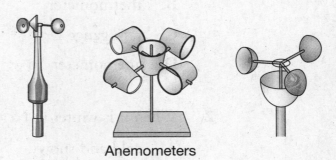

Anemometers

Precipitation can be measured in several ways. A **rain gauge** can measure the amount of rain that falls. This tool is a can or tube with an opening on top. Rain falls into the rain gauge. A scale on the side measures how much rain has collected in it.

Rain gauge

To measure snowfall, scientists use a tool similar to a rain gauge. Sometimes scientists collect snow in a container as it falls. Then they melt the snow and measure the amount of water.

DISCUSSION QUESTION
What do you think the weather will be tomorrow? How did you come up with your prediction?

LESSON REVIEW

1. Which of these tools measures wind speed?

   A. wind vane

   B. thermometer

   C. rain gauge

   D. anemometer

2. When it is winter in New York, it is MOST LIKELY to be

   A. cold and snowy.

   B. hot and windy.

   C. mild and rainy.

   D. warm and dry.

3. Study the bar graph below.

   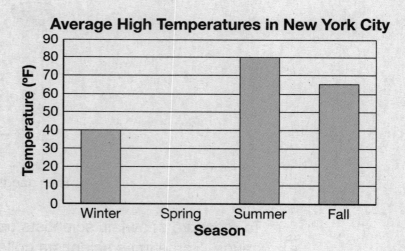

   The bar labeled "Spring" would MOST LIKELY reach to about

   A. 40 degrees.

   B. 60 degrees.

   C. 80 degrees.

   D. 20 degrees.

Lesson 19: The Water Cycle

PS2.1c; LE6.2c

Getting the Idea

Key Words
water cycle
evaporation
condensation
precipitation
groundwater
runoff

Water on Earth is constantly moving. Water moves from the land and oceans into the air and back again. This movement is called the **water cycle**.

Evaporation

The sun provides the energy that powers the water cycle. As the sun shines on the oceans and other bodies of water, some water near the surface changes into water vapor. *Water vapor* is water in a gas form. The change of a liquid to a gas is called **evaporation**.

Most water vapor moves into the air from the oceans. Water is also released as water vapor from the leaves of plants and from animals.

Condensation

When there are many clouds in the sky, it may mean that rain or snow is coming. A cloud is a mass of water droplets. High in the sky, water vapor cools and forms tiny droplets of water. This process is called condensation. **Condensation** is the change of a gas to a liquid.

Condensation does not happen only in clouds. You may have noticed that when you leave a cold drink out on a hot day, the outside of the glass feels wet. This happens because water vapor in the air condenses on the cold surface of the glass.

> **Did You Know?**
>
> Trees sweat! Up to 1,680 gallons of water evaporate from a large oak tree each day.

The water droplets in clouds stick together to form bigger drops. When the drops get heavy enough, they fall to the ground as precipitation. **Precipitation** is any form of water that falls from clouds, such as rain. In cold weather, precipitation often falls as snow or sleet. Hail is another form of precipitation.

The diagram below illustrates the movement of water through Earth's air, land, and water. The movement of water is a continuous cycle.

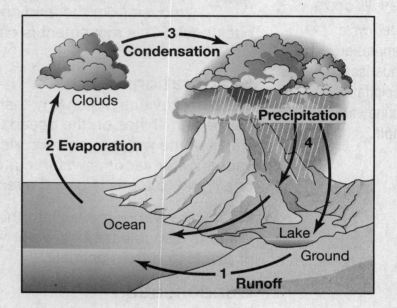

Water on Land

When water falls to Earth as rain, some of the water soaks into the ground. Water that is below Earth's surface is called **groundwater**. Much of the water that reaches Earth as precipitation flows over the surface of the land. This water is called **runoff**. Runoff also happens when snow melts in the spring. Runoff flows downhill, from high ground to low ground. The water travels from mountaintops into rivers that eventually empty into oceans.

Some water from rivers and oceans evaporates and returns to the air. The water that returns to the air from rivers, oceans, lakes, ponds, and living things allows the water cycle to keep repeating.

Lesson 19: The Water Cycle

The Water Cycle and Weather

Now that you know how the water cycle works, how do you think the water cycle relates to weather? When water evaporates, the humidity of the air increases. *Humidity* is the amount of water vapor in the air. Humidity, along with the temperature of the air, affects weather. Once water vapor condenses in the air, the water falls to earth as rain, snow, sleet, or hail. The water cycle is a very important part of understanding weather.

DISCUSSION QUESTION

How does water move through the water cycle where you live?

LESSON REVIEW

1. In which of these processes does water vapor become water droplets?

    A. evaporation

    B. precipitation

    C. runoff

    D. condensation

2. Rain, sleet, snow, and hail are all examples of

    A. precipitation.

    B. evaporation.

    C. condensation.

    D. weathering.

3. Which of the following does NOT describe how water moves?

    A. Water falls from the sky as rain. It stays where it falls until it evaporates.

    B. Water falls from the sky as snow. As the snow melts, the water runs uphill.

    C. Water falls from the sky as rain. The water then runs downhill.

    D. Water falls from the sky as hail. The hail melts and is absorbed into the ground.

4. Water does NOT enter the air from

    A. lakes.

    B. rivers.

    C. groundwater.

    D. oceans.

Lesson 20: How Earth's Land Changes

PS2.1d

Getting the Idea

Right at this moment, Earth's land features are changing. You do not see these changes because they usually happen very slowly. Wind and water work together to wear down the land in some places and build it up in other places. In this lesson, you will learn about how this happens.

Key Words
landform
weathering
erosion
deposition
sediment
soil
humus
nutrient
topsoil

Weathering

Earth's land surface is made up mostly of rock. Mountains and other surface features of the land, called **landforms**, are made up of rock. Rock is hard, but wind and water can slowly break rock apart. The breaking down of rock on Earth's surface into smaller pieces is called **weathering**.

Wind and water can carry sand and other small pieces of rock. You know that you can smooth a piece of wood with sandpaper. Blowing sand acts like sandpaper. It scrapes and wears away the surface of rock. Sand and pebbles carried by moving water also weather rock.

Rain causes weathering, too. Rainwater can seep into cracks in rock. If the temperature gets cold enough, the water freezes. When water freezes, it expands, or takes up more space. The ice in the cracks pushes against the rock. Over time, repeated freezing and thawing of water may split the rock apart. Plants also cause weathering. The roots of plants can grow into cracks in rock. As the roots become larger, they break the rock into smaller pieces.

Did You Know?

New York's Taughannock Falls is the highest free-falling waterfall in the northeastern United States. At 215 feet, it is 33 feet taller than Niagara Falls. The nearby gorge formed as Taughannock Creek wore through hundreds of feet of rock.

Wind, water, and plants work together to weather rock. Over thousands of years, weathering shapes rock into different formations. Some rock formations, like the one shown below, look like arches. Others look like tall columns. Weathering also helped form the canyon shown below.

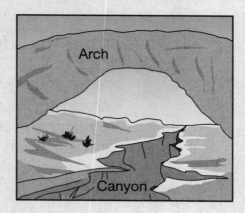

Erosion and Deposition

Weathering over long periods of time breaks rock into smaller and smaller pieces. Some of the pieces stay in place, but many get moved away. The process by which weathered rock is picked up and moved to a new place is called **erosion**.

Blowing wind picks up and carries sand. Strong winds can bounce and roll larger pieces of rock along the ground. Flowing rivers pick up sand and pebbles. Rivers may carry this material over long distances. Ocean waves also pick up and move sand and other pieces of weathered rock.

Gravity is a force that pulls everything down toward the center of Earth. Gravity causes erosion by pulling pieces of rock downhill.

Glaciers are huge sheets of ice that move slowly over land. As glaciers move, they pick up pieces of rock. A glacier can carry boulders the size of cars.

Glaciers also leave rocks behind. **Deposition** is the process by which eroded rock is dropped in new places. The material that is deposited is called **sediment**. When wind and water slow down or stop, they drop pieces of rock as sediment.

Soil

A material called **soil** covers most of Earth's land surface. Sediment is the main material in soil. Soil also contains the remains of dead plants and animals. Over time, these remains break down into small bits. These bits make up a dark-colored material called **humus**. Humus helps soil hold water.

Soil also contains nutrients. A **nutrient** is something that living things need to live and grow. Most plants grow best in soil that contains plenty of humus. Earthworms and other animals add nutrients to soil, too. The nutrients are in the wastes that the animals leave in the soil.

Layers of Soil

The soil that covers Earth's land has different layers. The top layer is called **topsoil**. Most plants grow in this soil layer. Topsoil has lots of humus in it. Insects, worms, and other animals make their homes in topsoil.

The next layer is subsoil. Subsoil is mostly rock particles, with very little humus. Roots of trees grow down into subsoil. Under the subsoil is a layer of weathered rock. The rock is broken up, but the pieces are bigger than soil particles. Under the weathered rock is solid rock, called bedrock. The picture below shows the layers of soil on top of bedrock.

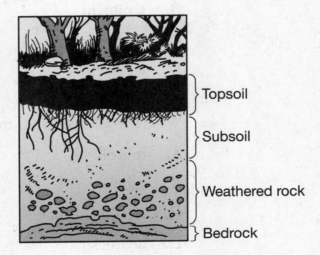

Test Tips...

Some questions ask about the definition of a word. Try to define the word yourself before you look at the possible answers. Then look for your definition among the choices.

DISCUSSION QUESTION
How are topsoil and subsoil alike? How are they different?

LESSON REVIEW

1. What is one way that water changes Earth's landforms?
    A. Water pulls weathered rock downhill.
    B. Water pushes weathered rock uphill.
    C. Water glues weathered rock together.
    D. Water holds weathered rock in place.

2. Which one is a definition of weathering?
    A. the movement of materials from one place to another
    B. the breaking down of rocks into smaller pieces
    C. the dropping of rock by wind or water
    D. the forming of a mountain

3. Which word names the process of picking up and moving weathered rock?
    A. deposition
    B. sediment
    C. erosion
    D. weathering

4. Where would you find the most plant roots and animals?
    A. topsoil
    B. subsoil
    C. weathered rock
    D. bedrock

Lesson 21: Extreme Natural Events

PS2.1e

Getting the Idea

Key Words
- hurricane
- flood
- thunderstorm
- tornado
- volcano
- earthquake
- wildfire

In Lesson 20, you learned about processes that change Earth's land slowly. Other changes happen fast. Hurricanes, floods, earthquakes, and volcanoes can change Earth's surface very quickly. It often seems that these fast changes are harmful to all living things. But they can sometimes help living things, too.

Hurricanes

Severe weather, such as hurricanes or tornadoes, can be very dangerous to people and other living things. A **hurricane** is a huge rainstorm that brings very strong winds with it. Hurricanes start over the ocean. If they come to shore, they can damage buildings and trees. A hurricane can stay in an area for several hours or even a few days. You should stay away from windows during a hurricane. The winds can blow so hard that the glass can break.

Floods

A **flood** is an overflow of water onto land that is usually dry. Floods often happen during a hurricane, when several inches of rain fall in a few hours. Floods also can happen in the spring, when snow on mountains melts. The water from the melted snow flows downhill into streams and rivers. When there is more water than the river can hold, the water floods the land near the river.

Floods can damage the homes of people and other living things. But not all the effects of floods are bad. Recall that flowing water carries sediments. A flood carries sediments and nutrients over the land. When the water from the flood drains away, it leaves the sediment and nutrients behind. These materials make the soil better for growing plants. A flood can improve farmland.

Thunderstorms

A **thunderstorm** is a heavy rainstorm that includes lightning and thunder. It is important to stay inside during a thunderstorm if you can, so you will not be hit by lightning. Lightning strikes very tall objects. If you cannot get inside during a thunderstorm, you should stay away from tall trees and other tall objects that lightning might hit.

Tornadoes

A **tornado** is a whirling wind that looks like a dark cloud in the shape of a funnel. A tornado spins very fast. It picks up objects as it moves across the land or water. A tornado can completely destroy one house, but leave the house next door standing. Tornadoes form in thunderstorms. If a tornado is in the area, you should go into a basement or a part of a building that has no windows.

"Tornado Alley" is a name for the area of the United States that gets many tornadoes each year. This area includes eastern Nebraska, Kansas, Oklahoma, South Dakota, northern Texas, and eastern Colorado. The flat land in the Great Plains allows cold, dry air from Canada to meet warm, moist air from the Gulf of Mexico. A large number of tornadoes form when these two air masses meet.

Lesson 21: **Extreme Natural Events**

Did You Know?

The islands that make up the state of Hawaii formed from volcanoes that erupted on the ocean floor. Each time they erupted, rock material built up. Today, the volcanoes stick up above the surface of the ocean, forming the islands.

Volcanoes

Earth's land surface is solid. But deep beneath the surface, there is hot, melted rock. A **volcano** is an opening in Earth's surface that lets melted rock and gases escape. When heat and pressure build up in the melted rock, the volcano erupts. Ash and gases shoot into the air, and melted rock flows out of the opening. The diagram below shows an erupting volcano.

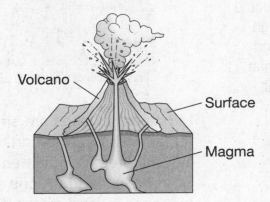

A volcanic eruption causes changes that happen in minutes, hours, or days. The explosion may blow off the top of the volcano itself. Ash, mud, and melted rock cover the land around the volcano. As the melted rock cools, it hardens into solid rock.

An eruption causes damage to the area around the volcano. As the melted rock moves downhill, it knocks down trees and kills small plants and animals. Ash can cover cars and houses. Some time after an eruption, plants begin to grow in the ash. The volcano helps some plants because the ash adds nutrients to the soil. That is one reason why people all over the world live near volcanoes, even though an eruption could be dangerous.

Earthquakes

An **earthquake** is the shaking that happens when large areas of rock under Earth's surface break and slip. Pressure that builds up inside Earth causes earthquakes. When the pressure becomes too great, rock breaks and moves suddenly. A place where pressure builds and where rock can break or slip is called a fault.

> **Did You Know?**
> New York State has over 70,000 miles of rivers and streams. If they were all end to end, that would be long enough to wrap around Earth more than two and a half times!

An earthquake releases energy. The energy makes the ground shake. This shaking can cause bridges and buildings to fall down. The diagram below shows how rock slips and releases energy.

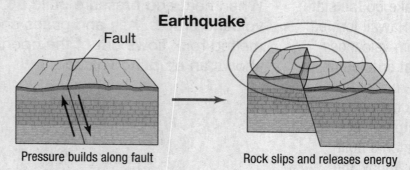

Most earthquakes are small and do not change Earth's surface. But strong earthquakes can change the land in just minutes. Sometimes land along a fault slips sideways. Sometimes land on one side of a fault is pushed up higher than land on the other side, forming steep cliffs. Sometimes the land pulls apart, forming a deep trench.

Wildfires

A **wildfire** is a fire in a natural area. It can be in a forest or a grassland. Lightning causes many wildfires. Sometimes fires can be caused by accidents, such as when a campfire gets out of control or someone drops a burning match. Wildfires usually start when the area is dry.

Once a wildfire starts, it can burn huge areas. It kills many plants. It kills animals that are not able to move away. A wildfire can spread to areas where people live. When this happens, people must leave their houses in order to stay safe.

DISCUSSION QUESTION

How do volcanic eruptions and earthquakes affect people and other living things?

LESSON REVIEW

1. What is the weather like during a hurricane?
    A. snowy and cold
    B. warm and dry
    C. windy and wet
    D. snowy and warm

2. Where do tornadoes come from?
    A. thunderstorms
    B. volcanoes
    C. blizzards
    D. earthquakes

3. A place where rock breaks and slips is called a
    A. plate.
    B. volcano.
    C. fault.
    D. wave.

4. Which of these causes melted rock to flow onto Earth's surface?
    A. earthquake
    B. volcanic eruption
    C. deposition
    D. landslide

Chapter 5 Review

1. The condition of the outside air at a certain time and place is called

    A precipitation
    B weather
    C condensation
    D evaporation

2. Which weather instrument should you use to find out how fast the wind is blowing?

    A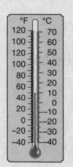

    B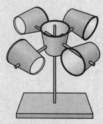

    C

    D

3. Which of these is *not* a form of precipitation?

    A snow
    B sleet
    C runoff
    D rain

4 Besides snow, which form of precipitation are you *most likely* to see in New York in winter?

  A rain
  B hail
  C clouds
  D sleet

5 What are clouds made of?

  A water droplets
  B rain and hail
  C snow and sleet
  D Scientists do not know.

6 What process moves water from Earth's surface into the air?

  A condensation
  B deposition
  C evaporation
  D weathering

7 What is the source of energy that keeps the water cycle going?

A wind
B rain
C the sun
D heat inside Earth

8 Condensation causes which of these to form?

A air
B clouds and precipitation
C runoff
D groundwater

9 What is weathering?

A the movement of materials from one place to another
B the breaking up of rock into smaller pieces
C the building up of sediments
D the release of melted rock from a mountain

10 What is the process by which sediment is dropped by wind or water?

A erosion
B weathering
C deposition
D condensation

11 The diagram below shows a rock arch over a canyon.

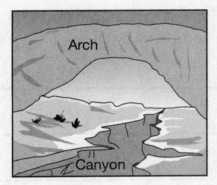

Which of these formed the arch?

A erosion by a glacier
B weathering by wind
C erosion by water
D deposition by wind

12 Which of these can add nutrients to the soil?

A volcano
B earthquake
C hurricane
D tornado

13 Which of these can damage or destroy a house by shaking it?

    A volcano
    B earthquake
    C hurricane
    D tornado

14 Name **one** kind of extreme weather condition that can cause a flood.

_____

15 Use terms from this chapter to describe today's weather.

_____

_____

_____

CHAPTER 6
Matter

Lesson 22 Physical Properties of Matter

PS3.1a, b, c, e; PS3.2a; CT4

Getting the Idea

Have you ever played guessing games? Maybe the game started when someone said, "I'm thinking of something that is red." You might have asked if the thing was big or small. Maybe you asked if it was rough or smooth. You might have asked if it was heavy or light. When you describe something, you talk about the features of the matter that makes up the object.

Key Words
matter
volume
mass
physical property
texture
flexible
conductive
state of matter

Matter

Look around you. Every object you see is made of matter. **Matter** is anything that has mass and takes up space. Your textbooks are made of matter. You know that they take up space because they fill the space in your backpack. Once your backpack is full, you cannot add another book. This happens because two objects cannot occupy the same space at the same time. An object's **volume** is the amount of space the object takes up.

The matter that makes up your textbooks has mass. **Mass** is the amount of matter that makes up an object. The mass of an object determines how much Earth's gravity pulls on it. You can tell that books have mass because you can feel how heavy your backpack gets when it is full.

Describing Matter

Each kind of matter has properties that you can use to describe it. A **physical property** is a feature of matter that you can observe with your senses or measure with a tool. Color, shape, and size are physical properties. So are taste and odor, or smell. Weight and temperature are physical properties you can measure. You will learn more about measuring physical properties in Lesson 24.

Lesson 22: **Physical Properties of Matter**

Properties can help identify different kinds of matter. For example, each kind of matter is a certain color. Rubies are red. Lemons are yellow. **Texture** is the way a surface feels, such as rough or smooth. You observe texture by touching things. Different kinds of matter have different textures. Substances such as glass and metal are smooth. Bricks are rough.

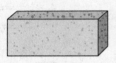
Brick

Eyeglasses

Some kinds of matter are more flexible than others. Something is **flexible** if you can bend it. For example, you can bend an aluminum soda can because aluminum is flexible. You cannot bend a drinking glass, even if it is very thin, because glass is not flexible. If you try to bend a glass, it will break.

Have you ever tried moving a refrigerator magnet to another place? If you put the magnet on something made of wood or plastic, the magnet did not stick. Only a few kinds of matter are attracted to a magnet. The magnet sticks to the door of the refrigerator because the door is made of steel. Steel is attracted to a magnet. A magnet does not attract plastic, wood, or glass.

Some kinds of matter carry electrical energy. Anything that can carry electrical energy is **conductive**. Copper, silver, and some other metals are conductive. People use metals such as copper to make wires. Wires carry electrical energy through a house. You can tell if a substance is conductive if it allows electrical energy to pass through it. You will learn more about electrical energy and conductive materials in Lesson 29.

Did You Know?

The only rock with a lower density than water is pumice. If you place a piece of pumice in a bowl of water, it floats!

Another physical property is whether a substance dissolves in water. Sugar and salt both dissolve in water. If you stir sugar into water, it disappear seems to into the water. It is still there, but you cannot see it anymore. Stainless steel does not dissolve in water. If you put sugar into a glass of water and stir it with a stainless steel spoon, the sugar will dissolve. The spoon will not.

Sugar dissolves in water, but stainless steel does not.

Whether an object sinks or floats is also a physical property. This property depends partly on the material an object is made of. Wood floats in water. So does a ball filled with air. Metals usually sink.

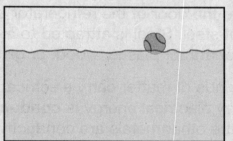 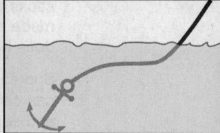

States of Matter

Matter has different physical forms, called **states of matter**. Most matter around you is in one of three states—solid, liquid, or gas. A material's state of matter is a physical property. You will learn more about states of matter in Lesson 23.

Lesson 22: **Physical Properties of Matter**

Test Tips...

Whenever a picture is included with a question, study it very carefully. Make sure you understand what the picture shows before you try to answer the question.

DISCUSSION QUESTION

What is a physical property of matter? Choose an object near you and describe as many of its physical properties as you can.

LESSON REVIEW

1. Which one describes a physical property of wood?
    A. floating in water
    B. sinking in water
    C. sticking to a magnet
    D. feeling greasy

2. All matter
    A. is conductive.
    B. is solid.
    C. has mass and volume.
    D. is flexible.

3. Something that is conductive is able to carry
    A. magnetism.
    B. texture.
    C. electrical energy.
    D. mass.

Lesson 23 Solids, Liquids, and Gases

PS2.1c; PS3.2a, b, c

Getting the Idea

Key Words
state of matter
solid
liquid
gas
physical change
change of state

Think about water. You know that it can be a liquid. Water can also be a gas in the air. And when water is frozen, it is a solid. In this lesson, you will learn more about what it means for something to be a solid, a liquid, or a gas.

All matter is made up of tiny particles. The particles are too small to see with your eyes. Scientists see these particles only by using very powerful microscopes. Recall from Lesson 22 that a **state of matter** is the form that matter takes. The particles of solids, liquids, and gases are arranged differently. Each of these three states of matter has different qualities.

Solid State

A **solid** is matter that has a definite shape. It also has a definite volume. Recall that volume is the amount of space an object takes up. A solid generally keeps its shape unless something breaks or bends it. A solid takes up a set amount of space. It does not spread out. The particles in a solid are packed tightly together. The picture below models what the particles in a solid might be like. The particles have energy. They vibrate, or move back and forth quickly. But they do not move away from their positions.

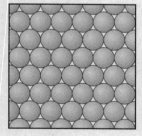

Lesson 23: Solids, Liquids, and Gases

Did You Know?

If you could magnify one glass of water until the glass of water was as big as Earth, each water particle would only be about the size of a baseball. Billions of baseball-sized particles jumbled together would represent the particles of a single glass of water.

Liquid State

A **liquid** is matter that has a definite volume but not a definite shape. A liquid takes the shape of its container. If you pour a liquid from a glass into a bowl, the shape of the liquid changes.

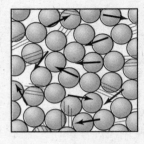

The particles of a liquid are not fixed in place like those of a solid. The particles of a liquid can slide past each other. The picture above models the particles of a liquid.

Gas State

A **gas** is matter that does not have a definite shape or a definite volume. A gas spreads out in all directions. Helium inside a balloon is a gas. The air you breathe is made up of gases. The particles of a gas are very far apart. They move around easily and bounce off each other. The picture at the right models the particles of a gas.

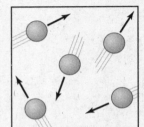

You can use state of matter to describe objects and materials. An ice cube and water in a pond are both water. But the ice cube is a solid, and the pond water is a liquid.

State of Matter

Solid	Liquid	Gas
Fixed volume	Fixed volume	Volume can change
Fixed shape	Shape can change	Shape can change
Examples: rock, ball, chair, spoon	Examples: water, oil, lemonade, tea	Examples: air, carbon dioxide, helium in a balloon

Duplicating any part of this book is prohibited by law.

Changes in State

You can change the size, shape, or form of matter. For example, if you break a crayon into pieces, it changes its shape and size. These changes are called **physical changes**. In a physical change, the kind of matter stays the same. No new materials form.

You can observe many examples of matter changing from one form to another. If you put an ice cube on a plate on a hot, sunny day, the ice cube will melt. If you put a pan of water in a freezer, the water will turn to ice. A puddle of water on the sidewalk will dry up. The water will change to a gas and disappear into the air. A change of matter from one form to another is called a **change of state**.

When a solid changes state to a liquid, it melts. Ice is water in the solid form. When ice is heated, it melts, changing into a liquid. Heat energy causes the particles to speed up. As the particles move faster, they move more. The ice turns into liquid water.

Frozen ice　　　Liquid water

Think about a pot of water heated on the stove. As the liquid water becomes hotter, the particles in the liquid move faster. Bubbles of gas form in the liquid and rise to the surface. At the surface, the gas disappears into the air. This process is called *boiling*.

Liquid water can change to a gas without boiling. The liquid water in a puddle does not boil, but it changes to the gas called water vapor. As you learned in Lesson 19, the change of a liquid to a gas is called *evaporation*.

Lesson 23: Solids, Liquids, and Gases

Did You Know?

Oxygen becomes a liquid at about minus 190 degrees Celsius.

Water vapor can change back into liquid water. This process is called *condensation*. Imagine it is a hot, sunny day. You hold a cold glass of water filled with ice cubes. You notice that the outside of the glass has drops of water on it. Water vapor from the air has condensed on the glass. When water vapor cools, its particles begin to move more slowly. As the particles move more slowly, they begin to pack more closely together. The water vapor, a gas, condenses. It turns back into liquid water.

Liquid water can also change to solid ice. When liquid water cools, the particles of the liquid water move more slowly. As the liquid loses more heat, it begins to change state. Soon, the liquid water freezes. The water changes into its solid state, ice.

The picture below summarizes the changes of state of water.

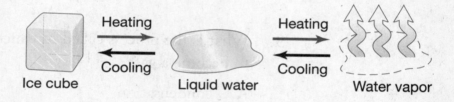

DISCUSSION QUESTION

Look at the drawings below. How would you know which ones represent a solid, liquid, or gas?

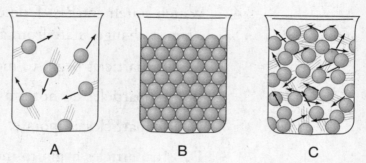

Test Tips...

Some test questions may ask you for the best choice. In that case, some of the wrong answers may look correct. You must read all the answers carefully to find the one that answers the question best.

LESSON REVIEW

1. Which is an example of a liquid?
   A. fruit juice
   B. snow
   C. wind
   D. string beans

2. A form of matter has no fixed shape or volume. The particles are very far apart. What form of matter is this?
   A. solid
   B. liquid
   C. gas
   D. ice

3. What is the process called in which a liquid changes state to become a gas?
   A. condensation
   B. evaporation
   C. melting
   D. freezing

4. Which statement BEST describes what happens to matter when it changes state from a liquid to a solid?
   A. Its particles begin to move faster.
   B. Its particles do not move at all.
   C. Its particles evaporate.
   D. Its particles begin to move more slowly.

Lesson 24: Measuring Properties

M1.1a, b, c; M3.1a; PS3.1c, e, g

Getting the Idea

Some physical properties must be measured with a tool. You must choose the right tool for the kind of measurement you want to make. You must also know how to use the tool correctly.

Key Words
- area
- volume
- mass
- weight

Measuring Length and Width

You can use a ruler to measure how long or wide something is. One kind of ruler is a meterstick. A meterstick is one meter long. It is divided into 100 centimeters. A meterstick is easy to use. First, make sure the end of the stick, or the 0 mark, is exactly at one edge of an object. Line up the meterstick with the other end. Then read the number that lines up with other end.

Measuring Area

An object's **area** is the amount of space its surface covers. You can use a ruler to find area, but you must also do some math. Suppose you want to set up a fish tank in your room. The tank must fit on a small table. Before you buy a tank, you must find the area of the tabletop. First, measure the length and width of the tabletop.

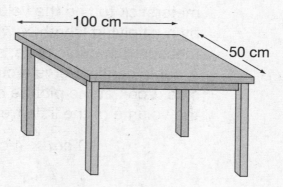

The tabletop is 100 centimeters long and 50 centimeters wide. To find its area, multiply its length times its width. Area is always given in square units, such as square centimeters (cm^2). A square centimeter is a square that measures 1 cm on each side.

$$100 \text{ cm} \times 50 \text{ cm} = 5000 \text{ cm}^2$$

Suppose your fish tank is 80 centimeters long and 40 centimeters wide. What is the area of the fish tank?

$$80 \text{ cm} \times 40 \text{ cm} = 3200 \text{ cm}^2$$

You see that the tank will fit on your table. The tank has a smaller area than the top of the table—3200 is less than 5000.

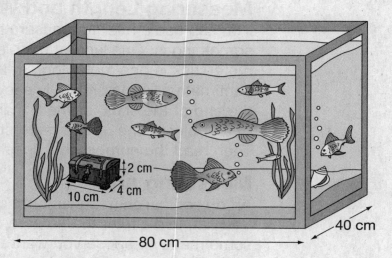

Measuring Volume

Volume is the amount of space that an object or material takes up. You can make measurements with a ruler or meterstick to find the volume of a solid. You find volume by multiplying length × width × height. Volume is often measured in cubic units, such as cubic centimeters (cm^3). A cubic centimeter is a cube that measures 1 cm on each side. Look at the picture of the fish tank above. What is the volume of the little treasure chest inside the fish tank?

$$10 \text{ cm} \times 4 \text{ cm} \times 2 \text{ cm} = 80 \text{ cm}^3$$

You can measure the volume of a liquid with a measuring cup or a graduated cylinder. First, pour the liquid into the cylinder. Find the line that matches the level of the liquid. Make sure that your eyes are at the same level as the marks on the cylinder. Figure out where the liquid and the lines meet.

Look at the drawings below. The graduated cylinder and the measuring cup both contain 50 milliliters of liquid. One milliliter takes up the same space as one cubic centimeter.

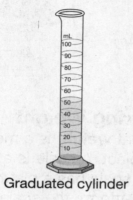

Graduated cylinder   Measuring cup

Measuring Temperature

A thermometer is used to measure temperature, or how warm something is. You learned about thermometers and temperature in Lesson 18. You can use a thermometer to measure the temperature of the air or another material. The temperature of an object changes depending on the conditions around it. For example, a car door feels hot when you touch it in summer. It feels cold when you touch it in winter.

Measuring Mass

Recall that **mass** is the amount of matter an object contains. A pan balance is used to measure mass. To use a balance, put an object on the left pan. Then add mass weights to the right pan until the pointer shows that the balance is level. Add up the masses of all the mass weights to find the mass of the object.

Measuring Weight

Recall that **weight** is a measure of the pull of gravity on an object. A scale is a tool for measuring weight. A digital scale has a platform. When you put an object on the platform, the scale measures how much the object is pulled down by gravity. Wait until the numbers stop changing before you record the measurement.

A spring scale has a hook attached to a spring. When you hang an object from the hook, it stretches the spring. A pointer attached to the spring moves. You can tell the weight of the object by looking at the number by the pointer.

Digital scale

Spring scale

Lesson 24: **Measuring Properties**

DISCUSSION QUESTION
Choose three tools you read about in this lesson. Tell how you might use each one to measure different objects around you.

LESSON REVIEW

1. Which tool would you use to measure the width of your classroom?

    A. graduated cylinder

    B. meterstick

    C. spring scale

    D. balance

2. A piece of note paper is 5 cm long and 4 cm wide. What is its area?

    A. 20 cm

    B. 9 cm^2

    C. 1 cm

    D. 20 cm^2

3. What can you measure with a graduated cylinder?

    A. weight

    B. length

    C. volume

    D. temperature

4. What can you use to measure the mass of an object?

    A. spring scale

    B. digital scale

    C. balance

    D. ruler

Lesson 25 Classifying Objects

S2.3a; PS3.1f

Getting the Idea

When you go to a library, you might find books about animals on one shelf and books about famous people on another. The books are organized into groups by topic. In much the same way, scientists organize objects into groups based on how they are alike or different. Rocks, minerals, and living organisms are often organized into groups.

Key Word
classify

Classifying

When you **classify** objects, you organize a large group of items into smaller groups, or categories. To classify objects into smaller groups, you need to use a characteristic or trait to compare the objects.

For example, suppose you want to place your clothes into groups in your closet. You might separate the clothes by whether you wear them in summer or in winter. Or, you might want to separate them by tops (shirts and jackets) and bottoms (shorts and pants.) You might even separate them by whether you wear them at home or at school. You can choose different characteristics. Once you decide, you compare the clothes and put them into groups.

Often, you will need to separate the smaller groups into even smaller groups. Think about your clothes again. Suppose you separated your clothes by tops and bottoms. Now think about how you can divide the tops into even smaller groups. Perhaps you will place the tops with short sleeves in one group and the tops with long sleeves in another group.

Lesson 25: **Classifying Objects**

Sorting by Properties

Scientists follow the same kinds of processes when they classify objects. They choose qualities to help make groups. Suppose you had all the items shown below. Think of what you read in Lesson 22 about properties of matter. Do any of these items share properties?

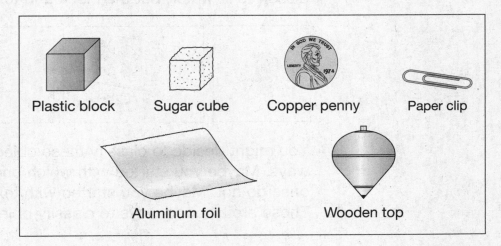

First, you might test which objects are conductive. You could classify them by that property. If you did this, you would find that three of these objects are conductive.

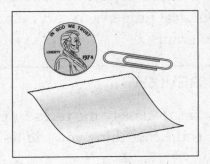
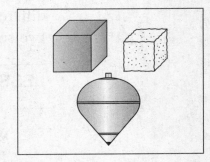

To further classify the objects in the "Conductive" box, you might look at color. The penny is copper colored. Steel and aluminum look silvery in color.

Now think about the objects that are not conductive. What happens if you put them in water? The sugar cube dissolves in water, but the block and top do not.

 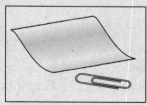 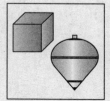

You might decide to classify these objects in different ways. Maybe you started with which ones float and which ones do not. Maybe you started with texture or flexibility. Those are all good ways to classify objects.

DISCUSSION QUESTION

Gather 12 small objects in your classroom. Use a physical property to sort the objects into two groups. Then use a different physical property to sort each of those groups into two smaller groups.

LESSON REVIEW

1. You want to classify the rocks in your rock collection. What is the first thing you should do?

    A. put the rocks into many piles

    B. make a list of the properties you observe

    C. throw out all the large rocks

    D. count the rocks

Lesson 25: **Classifying Objects**

Test Tips...

Stay positive. There is no reason to get upset if you do not know the answer to a question right away. Just move on and come back to it later.

2. A student has six rocks. The rocks all have different sizes and shapes, but some have the same colors. Which of the following properties could the student use to classify the rocks?

   A. shape
   B. size
   C. volume
   D. color

3. A student wants to classify these buttons by color. Which two buttons can be classified together?

   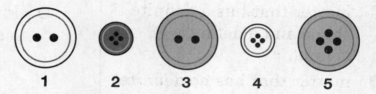

   A. 1 and 3
   B. 2 and 4
   C. 1 and 4
   D. 4 and 5

Review

1. Which of the following **best** describes a gas?

   A matter that has a definite shape but not a definite volume

   B matter that has a definite volume but not a definite shape

   C matter that has a definite shape and a definite volume

   D matter that has no definite shape and no definite volume

2. What happens to a scoop of ice cream when heat is added to it?

   A It melts.
   B It freezes.
   C It evaporates.
   D It condenses.

3. What happens when matter condenses?

   A It changes from a gas to a liquid.
   B It changes from a liquid to a gas.
   C It changes from a solid to a liquid.
   D It changes from a liquid to a solid.

4. What is matter?

   A anything that has mass and weight
   B anything that has color and texture
   C anything that has weight and color
   D anything that has mass and takes up space

5 Which property of matter would you observe by using your sense of touch?

A color
B texture
C taste
D shape

6 How would you calculate the volume of a block of wood?

A multiply length times width
B multiply length times width times height
C divide length by width
D divide mass by volume

7 Which tool can you use to find the weight of a rock?

A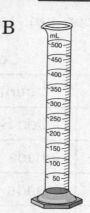

B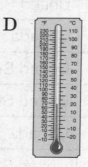

C

D

8 Which property of an object can change depending on the conditions around the object?

A magnetism
B texture
C color
D temperature

9 A floor is 4 meters wide and 6 meters long. What is the area of the floor?

_____

10 The table below contains information about coins.

Coin Collection

Coin	Country	Color	Name
A	United States	copper	penny
B	Canada	silvery	dime
C	Canada	copper	penny
D	Canada	copper	loonie
E	United States	silvery	nickel
F	Canada	silvery	quarter
G	United States	silvery	dime
H	United States	golden	dollar
I	Canada	silvery	nickel
J	United States	silvery	quarter

Sort the coins into **two** groups. Explain how you classified them.

(1) _____

(2) _____

I classified the coins by _____.

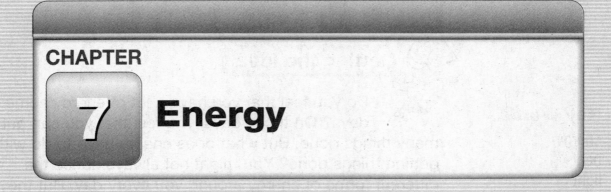

CHAPTER 7: Energy

Lesson 26 Forms of Energy

PS4.1a, b; PS4.1g

Getting the Idea

Do you feel that you have a lot of energy some days? On those days, you feel that you can get many things done. But what does energy have to do with getting things done? You might not always notice the different forms of energy around you every day. But they do help get things done. In this lesson, you will learn about some different forms of energy.

Key Words
energy
heat
light
sound
mechanical energy
electrical energy
chemical energy

Heat, Light, and Sound

Energy is the ability to do work or cause change. In science, *work* means moving an object from one place to another. You will learn more about work in Lesson 34.

There are many different forms of energy. They include heat, light, and sound. They also include mechanical, electrical, and chemical energy.

You know that matter is made up of tiny moving particles. The particles have energy. **Heat** is energy flowing from the particles of one object to the particles of another object. Heat warms your hands, cooks food, and melts ice. Some of the many sources of heat are shown below. They include the sun, Earth's interior, friction, and fire. You will learn more about heat and friction in Lesson 27.

156

Did You Know?

Only 4 percent of the energy put out by a lightbulb is light. Most of the energy is heat.

Light is a form of energy that you can see. It moves from place to place as waves. The sun gives off light and is Earth's main source of light. Fire and lightbulbs also give off light. Light makes it possible for you to see. You will learn more about light in Lesson 28.

Sound is a form of energy that you can hear. Sound moves in waves. Sound is produced when objects move very quickly, or vibrate. Many kinds of sounds can be made by musical instruments. You will learn more about sound in Lesson 28.

Other Forms of Energy

Mechanical energy is the energy of moving objects. A turning windmill and a moving car have mechanical energy. You use mechanical energy to throw a ball or jump.

Electrical energy is the energy of moving electric charges. Lightning is a form of electrical energy. You use electrical energy to run lamps, fans, and other appliances. You get this energy from batteries or from electrical outlets in the walls. You will learn more about electrical energy in Lesson 29.

Chemical energy is energy stored in a material. Chemical energy is stored in foods, fuels, and in your body. When you eat an apple, your body gains chemical energy that was stored in the fruit. That is the energy your body uses to move, think, and grow.

People burn fuels such as oil and wood to get energy stored in those materials. A battery also stores chemical energy. That energy can light a flashlight or make a cell phone work.

Transferring Energy

Energy can be transferred from one place to another. Energy transfers can be very useful. A hot radiator transfers heat into a room. The objects in the room also get warmer through energy transfer.

Transfers of energy can also be harmful. You have been told not to touch a hot pan that just came out of the oven. This is because a hot pan can be dangerous. The pan can transfer heat your hand and give you a burn.

Heat flows from one object to another. If you wash your hands in warm water, heat moves from the water to your hands. You can feel the warmth from the water. If the water is too hot, that energy can hurt your hands. You will learn more about heat moving from one place or object to another in Lesson 27.

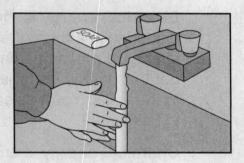

DISCUSSION QUESTION
What forms of energy did you use today? Explain your answer.

Test Tips...

Pay special attention to words such as *not* or *never* in a multiple-choice question or answer. These words are small, and you might not notice them. However, these words change the meaning of what you read.

LESSON REVIEW

1. Which one BEST describes energy?
   A. ability to cause change
   B. movement of an object
   C. fuel used to power a car
   D. movement of electric charges

2. Which one is the energy of moving objects?
   A. mechanical energy
   B. sound
   C. chemical energy
   D. light

3. Which one is a source of chemical energy?
   A. a hot stove
   B. a turning windmill
   C. a bright lightbulb
   D. food

4. Which one is NOT a source of heat?
   A. the sun
   B. ocean waves
   C. fire
   D. friction

Lesson 27 Heat and Temperature

M3.1a; PS4.1a, b, c, f, g

Getting the Idea

Think about a sunny summer day. A weather reporter on TV says that the temperature is 90°F. When you go outside, the air feels very warm. Soon your body feels hot. You are feeling heat from the sun. That energy went from the sun to your body. In this lesson you will learn how heat moves from one place to another.

Key Words
heat
chemical change
friction
conduction
conductor
insulator
temperature
thermometer

Sources of Heat

Heat is energy flowing from the particles of one object or material to the particles of another object or material. Heat from the sun warms Earth's air, water, and land. Without heat energy from the sun, Earth would not be warm enough for living things. But the sun is not the only source of heat on Earth.

Some chemical changes cause materials to give off heat. A **chemical change** is a change in matter in which new materials form. A chemical change takes place when wood burns. People also get heat by burning other fuels, such as oil and coal. A *fuel* is any material that people burn to get energy. Your body gets energy from chemical changes in food, and some of that energy warms your body.

Friction also produces heat. **Friction** is a force that acts when two things touch or rub together. If you rub your hands together quickly, you feel them get warmer. Friction changes some of the energy of your moving hands into heat.

How Heat Moves

Heat moves through solids, liquids, and gases. It also moves through empty space. Heat from the sun travels through space before it warms Earth's air.

One way that heat moves is by conduction. **Conduction** is the flow of heat between two objects that touch each other. Heat flows from a cup of hot chocolate to your hands by conduction. The hot liquid touches the cup, and the cup touches your hands. If you put a metal spoon in the cup, heat will flow to the spoon, too.

Heat always moves from a warmer object to a cooler one. Heat does not move from your hands to a cup of hot chocolate—unless your hands are even hotter than the cup. Think of when you hold a cold glass of water. In that case, heat moves from your hands to the glass.

Heat moves to your hand, your hand feels warm.

Heat moves to the glass, your hand feels cold.

Suppose you put a pot of water on the stove and turn up the burner. Soon the water begins to boil. You put a metal spoon in the water. You also put a wooden spoon in the water. Heat from the water moves to the two spoons. But only one spoon becomes too hot to touch. Which one?

The metal spoon is the one that is too hot to touch. Heat moves through some substances better than others. **Conductors** are materials that heat flows through easily.

Did You Know?

Wearing a hat when it is cold is good advice. A hat slows the movement of heat away from your head. Your head is more sensitive to changes in temperature than some other parts of your body. So weaping a hat makes you feel warmer.

Metals are good conductors. **Insulators** are materials that heat does not flow through easily. Plastic, wood, and glass are insulators. The plastic or wood handle on a cooking pot does not transfer heat from the pot.

Cloth is also a good insulator. Think about going outside on a cold day. You wear layers of clothes to keep warm. Layers of clothes keep warmer air close to your body. Heat does not pass to the air outside easily. That is what keeps you warmer.

Measuring Heat

Recall that **temperature** is a measure of how warm something is. When heat moves into an object, its temperature rises. When heat moves away from an object, its temperature gets lower.

As you learned in Lesson 18, a **thermometer** is a tool for measuring temperature. Marks on the thermometer show a scale. This scale is divided into units called degrees. The thermometer in the picture has two scales. They show the temperature in degrees Fahrenheit (°F) and degrees Celsius (°C).

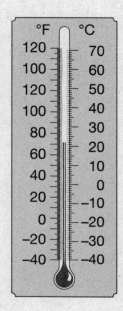

DISCUSSION QUESTION

Look at the pictures below. Classify all these objects as either conductors or insulators.

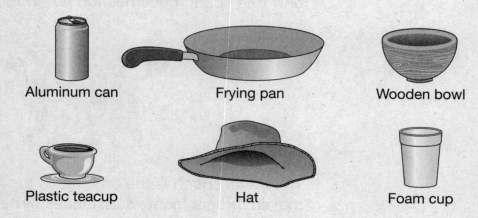

Aluminum can　　　Frying pan　　　Wooden bowl

Plastic teacup　　　Hat　　　Foam cup

LESSON REVIEW

1. A puddle of water on the sidewalk feels warm when you touch it. Which statement BEST explains why the water feels warm?

    A. Heat moved from the puddle to the sidewalk.

    B. Burning fuel warmed the puddle.

    C. Energy from the sun warmed the puddle.

    D. Friction from rainwater caused the puddle to get warm.

2. Which statement about heat is TRUE?

    A. Heat always moves from objects with lower temperatures to objects with higher temperatures.

    B. Heat always moves from objects with higher temperatures to objects with lower temperatures.

    C. Heat sometimes moves from objects with lower temperatures to objects with higher temperatures.

    D. Heat does not usually move.

3. A carpenter cuts a piece of wood with a saw. The carpenter moves the saw back and forth very fast. Afterward, the edge of the saw feels hot. What made the saw hot?

    A. melting   C. burning

    B. conduction   D. friction

4. Which one does NOT produce heat?

    A. a frying pan   C. the sun

    B. friction   D. fire

Lesson 28 Sound and Light

PS3.1g; PS4.1a, b, d, g

Getting the Idea

You use your senses to learn about your environment. Most people use their sight and hearing more than their other senses. Light makes it possible for you to see. Sound is what you can hear.

Key Words
sound
vibrate
light
reflect
absorb

Sound and Vibrations

When you hear a person talking, a dog barking, or a police siren, you hear sound. Some sounds are pleasant to hear, such as the sound of a song or of water in a stream. Some sounds are helpful, such as the bell that signals a fire drill at school. Some sounds may hurt you. Loud music or the roar of a jet plane can hurt your ears.

All sounds have certain things in common. **Sound** is a form of energy you can hear. It moves through matter as waves. All sounds are made by something vibrating. When something **vibrates**, it moves back and forth quickly.

Think about what happens when someone plucks guitar strings. The strings vibrate. Each time they move, the strings bump into nearby air particles. These particles bump against other particles. The energy moves through the air in waves. The drawing below models how sound waves move through the air.

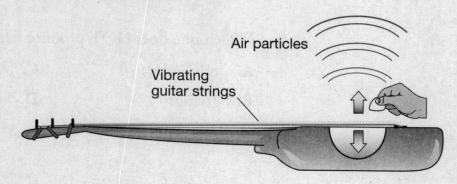

Lesson 28: Sound and Light

When sound waves reach your ears, the air in your ears vibrates. That makes your eardrum inside your ear vibrate. Signals travel to your brain, and your brain interprets the signals as sounds.

Did You Know?

Since there is no air in outer space, there is nothing there to carry sound from one place to another. Your ears would not hear anything without air to carry the sound waves.

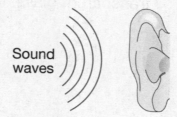

Light

Light is a form of energy that we can see. The sun, a burning candle, and a lightbulb all give off light. Light can travel through space. It does not need to travel through matter the way sound travels. Light travels in a straight line until it hits something. Depending on what object or material the light strikes, different things happen.

Light cannot pass through most objects and materials, such as metal or wood. When light strikes an object made of such materials, the object blocks some or all of the light. When an object blocks light, the object casts a shadow. You learned about shadows in Lesson 14. The camel below blocks the sunlight. That is why the camel casts a shadow on the ground beside it.

Did You Know?

Life on Earth depends on light from the sun. Green plants change light in a very important way. They use the energy of sunlight to make their own food.

Light can travel through some materials, such as glass, air, and water. These materials are transparent, or clear. When light strikes transparent materials, a shadow does not form. Most of the window shown below is made of transparent glass. It allows light to pass through it.

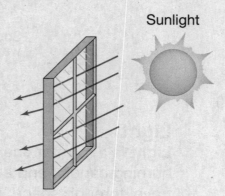

Sunlight

When light hits an object, the object **reflects** some of the light, or causes it to bounce back. Reflected light allows us to see things. In the picture below, arrows represent light. The light is reflected by two different surfaces. One surface is smooth, and the other is bumpy. Notice that the light waves move in different directions depending on the surface that reflects them.

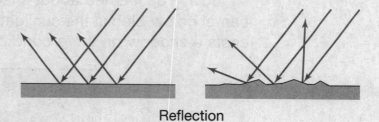

Reflection

Light and Heat

Just as an object reflects some light, the object also **absorbs**, or takes in, some light. Remember that light is a form of energy. The light that an object absorbs changes to a different form of energy—heat. This is why you get warm if you sit in the sun. You absorb the energy in sunlight. Light-colored materials tend to reflect more light than they absorb. Dark colored materials tend to absorb more light than they reflect. That is why people often wear lighter colors in summer than in winter. That is also why the dark surface of a road gets so much hotter than lighter-colored concrete.

Light and Color

Light has different colors. White light is made up of all the colors of the rainbow. When white light hits an object, some colors are absorbed. Other colors are reflected. The color your eyes and brain detect depends on the different colors of light that the object reflects.

DISCUSSION QUESTION

Study the picture of the drum below. Describe how the drum makes sound. How do you hear this sound? Explain the process.

LESSON REVIEW

1. What causes a sound?

    A. something vibrating

    B. a part of the ear

    C. energy from the sun

    D. energy stored in a material

2. When an object blocks light, the object

    A. changes color.

    B. helps light absorb things.

    C. makes light transparent.

    D. casts a shadow.

3. Which material can light pass through?

    A. metal

    B. concrete

    C. glass

    D. wood

4. Which form of energy can travel through empty space?

    A. light

    B. sound

    C. both light and sound

    D. neither light nor sound

Lesson 29 Electrical Circuits

PS4.1a, b, c, d, e, g

Getting the Idea

Key Words
electrical energy
circuit
conductor
insulator
closed circuit
open circuit

How many times a day do you switch on a lamp? Or turn on a television? Or use a computer? What do these things have in common? They all work using electrical energy. This energy moves through pathways called circuits.

How Electrical Energy Moves

Electrical energy is the energy of moving electric charges. Electrical energy can be used in many ways. It can be used to make light and heat. It can make things move, such as an electric fan. The fan will not turn unless electrical energy moves through it. To get electrical energy where you want it to go, you must give it a path to follow. A **circuit** is a path through which electrical energy can flow.

Electrical energy cannot move through all materials. A **conductor** is a material that electrical energy can flow through. Metals are good conductors. Many circuits use copper wire to carry electrical energy.

An **insulator** is a material that electrical energy cannot flow through. Plastic, glass, rubber, and wood are insulators. The wires in a circuit are usually covered with rubber or plastic to keep the electrical energy inside the wire.

The wires in the walls of your home form circuits. When you plug in a fan, the fan becomes part of one of those circuits. A circuit is made up of a source of electrical energy, wires, and devices that use electrical energy.

Did You Know?

New York's Brooklyn Bridge was the first bridge in the world to be lit using electricity.

An adult can help you make a simple circuit of your own. All you need is a battery, wire, and a lightbulb.

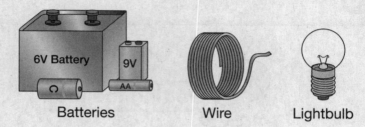

Batteries   Wire   Lightbulb

Batteries store energy. That energy can move through your circuit as electrical energy. The lightbulb can change electrical energy into light. The wire carries the electrical energy from the battery into the lightbulb.

Electrical energy will not flow in your circuit unless it is a complete loop. Electrical energy must be able to flow to and from the battery. It must also be able to flow through the bulb. So the wire must connect the battery to the light bulb on both sides. Your circuit might look like this.

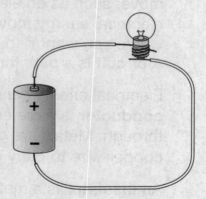

This circuit would work. But a circuit like this has a problem. The electrical energy would flow until the battery ran out of energy. You would not be able to turn the light off without breaking the circuit. That is why most circuits have a switch.

Lesson 29: Electrical Circuits

Open and Closed Circuits

If a circuit has a switch, you can turn the light on or off. The inside of the switch is made of metal, like the wire. The switch can carry electrical energy, too. When the switch is closed, the light is on. In this case, the circuit is closed. A **closed circuit** is a complete path for the flow of electrical energy. In a closed circuit, the energy from the battery flows through the wire and switch. This is what keeps the bulb lit. A closed circuit keeps electrical energy flowing.

When the switch is open, the light is off. In this case, the circuit is open. An **open circuit** is an incomplete path for the flow of electrical energy. The flow of electrical energy stops because the path is broken. When the circuit is open, the electrical energy cannot flow from the battery to the bulb. An open switch does not carry electrical energy from one part of the wire to the other part.

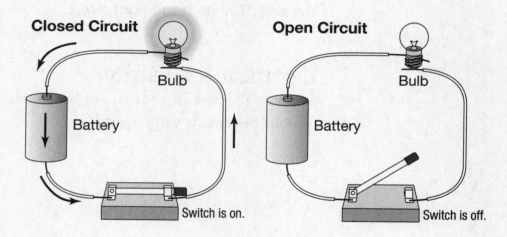

The arrows in the closed circuit above show the flow of electrical energy. Notice that the bulb is lit. No arrows are around the open circuit. Electrical energy cannot flow without a complete path. The lightbulb does not light.

Using Electrical Energy Carefully

Electrical energy is a very useful form of energy. But it can be dangerous, too. A circuit can become warm when electrical energy flows through it. If it gets too warm, this may be dangerous.

A circuit usually includes more than one electrical device. When you switch on lights in a room, more than one light might go on at once. But if too many devices are part of one circuit, the circuit can become hot. Too much energy flows through the circuit at once. This can start a fire. That is why you should not plug too many objects into the same wall outlet.

The wires in electrical devices are covered. This covering keeps them insulated. It protects you. If some of the covering comes off, you must never touch the bare wire. That could give you a dangerous shock. If you ever notice that the covering is coming off a wire, do not plug it into the wall. Tell an adult right away.

DISCUSSION QUESTION

Suppose the wire in your circuit broke. How might you fix it? Would the circuit work again?

Lesson 29: **Electrical Circuits**

LESSON REVIEW

1. Wire is needed to make a circuit because
   A. wire carries the electrical energy from one part to another.
   B. wire supplies the rest of the circuit with electrical energy.
   C. wire lights up when electrical energy reaches it.
   D. wire is a part of the circuit's battery.

2. The lightbulb in a circuit is off. It is MOST LIKELY that
   A. the lightbulb is in a closed circuit.
   B. the path of the circuit is broken.
   C. there are two lightbulbs in the circuit.
   D. there are two batteries in the circuit.

3. Which of these is a conductor of electrical energy?
   A. glass
   B. plastic
   C. copper
   D. rubber

4. You turn a circuit off and on using a
   A. wire.
   B. switch.
   C. battery.
   D. lightbulb.

Lesson

30 Energy Changing Form

M2.1b; PS4.1d; PS4.2a, b

Getting the Idea

Key Word

interact

Think of all the things around you that run on electrical energy. Electrical energy makes a toaster warm. It makes toy trains move. It makes a lightbulb light up. How does electrical energy do all these things? Electrical energy can change into other forms of energy. In this lesson, you will learn how energy changing from one form to another is part of your daily life.

Matter and Energy

One form of energy can change into another form of energy. These changes take place when energy and matter **interact**, or act on each other.

Light energy and matter interact all the time. When matter absorbs light, the matter gets warm. Light changes to heat. That heat can cause a puddle of water to evaporate. Darker objects absorb more light than lighter objects do. A darker object gets warmer than a lighter object when the light that is absorbed changes to heat.

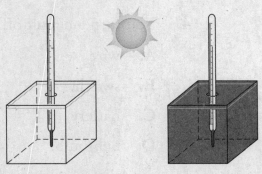

The white box is cooler because it reflects light.
The black box is warmer because it absorbs light.

Energy Changes Form

The table below shows some more ways in which energy changes from one form to another.

Some Energy Transformations	
Green plants change light into chemical energy. 	Burning fuels changes chemical energy into mechanical energy. 
A hot plate changes electrical energy to heat energy. 	A digital music player changes chemical energy (in the battery) to sound energy. 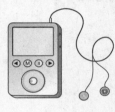
A fan changes electrical energy into mechanical energy. 	Solar panels on a satellite change light energy to electrical energy. 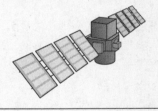
A lightbulb changes electrical energy to light energy. 	An animal's body changes chemical energy in food into mechanical energy. 

Test Tips...

A question may ask you to compare two things. That means you must decide how they are alike. A question may ask you to contrast two things. In that case, you must decide how they are different.

DISCUSSION QUESTION

Describe the energy changes that take place when you watch television.

LESSON REVIEW

1. Which example describes chemical energy changed into mechanical energy?

    A. fuel used to power a truck

    B. batteries used to turn on a flashlight

    C. wind used to power a windmill

    D. sunlight used to warm up a pool

2. A solar calculator turns light into

    A. sound.

    B. heat.

    C. electrical energy.

    D. mechanical energy.

3. What energy change happens when you clap your hands?

    A. sound to chemical

    B. mechanical to sound

    C. sound to mechanical

    D. electrical to sound

CHAPTER 7 Review

1. Which one has mechanical energy?

   A a baseball that has just been hit

   B a bat resting on a batter's shoulder

   C a baseball glove on a bench

   D a player standing at third base

2. Why does a metal spoon get hot when it is in a bowl of hot soup?

   A Electrical energy makes the spoon hot.

   B Chemical energy makes the spoon hot.

   C Heat moves from the soup to the spoon.

   D Heat moves from the spoon to the soup.

3. Which one is a conductor of heat?

   A wood

   B rubber

   C plastic

   D copper

4 Heat does *not* flow well through

A glass
B silver
C steel
D aluminum

5 Energy can be harmful if

A electrical energy is used to make light
B the sound from your headphones is too loud
C the sun makes puddles evaporate
D objects make shadows

6 Which of these lists the parts of a closed electrical circuit?

A wire, bulb, closed switch
B battery, bulb, open switch
C wire, battery, bulb, closed switch
D wire, battery, bulb, open switch

7 Which one is an example of heat that comes from friction?

    A The sun warms Earth.
    B You rub your feet on a rug.
    C A candle burns.
    D Water in a pan boils.

8 The diagram below shows an electrical circuit.

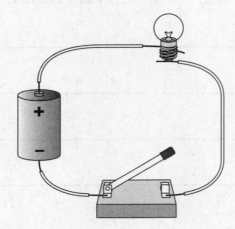

Which statement is *true* about this circuit?

    A The circuit is closed.
    B The circuit is open.
    C The bulb must be on.
    D Electrical energy is flowing through this circuit.

9 A battery lights the bulb inside a flashlight. Which choice *best* represents the changes in energy that take place?

    A heat energy → light energy → sound energy
    B light energy → mechanical energy → chemical energy
    C electrical energy → heat energy → light energy
    D chemical energy → electrical energy → light energy

10 When you turn on a television, electrical energy changes to

    A sound and mechanical energy
    B sound and chemical energy
    C light and sound energy
    D light and mechanical energy

11 On a hot summer day, you may feel warmer if you wear a black shirt rather than a white shirt. Explain **one** reason why this is true.

_____

_____

_____

_____

CHAPTER 8: Forces and Motion

Lesson 31 Motion

Getting the Idea

M1.1a; M2.1a; S2.3a; PS5.1a, b

Key Words
force
position
motion
direction
speed

How do you make something move? You apply a force. A **force** is a push or a pull. You push a shopping cart at the supermarket. You pull a wagon or a sled. You push or pull a door to open or close it. When you walk or run, you push your feet against the ground to move your body forward.

Position

A push or a pull can make an object start moving. When an object moves, its position changes. **Position** is the place where something is located. You describe an object's position by saying where it is compared with something else. Your book is in front of you. Your book may be behind another student. It may be next to someone else. Your book might be inside a room, or next to a wall. All these words describe position.

Changing an Object's Movement

Motion is an object's change in position over time. A force can change the direction of an object's motion. **Direction** is the path a moving object follows. You turn the handlebars to make a bike turn left or right as you pedal. When you hit a baseball, the bat pushes the ball away from you.

Another way that a force can act on an object is to stop its motion. You push on the brakes to stop a bicycle's motion. You pull on a shopping cart to stop it from rolling away. You will learn more about how forces change motion in Lesson 32.

Lesson 31: Motion

The chart below lists some ways to describe position and direction. You can use these words and others to tell how a force acts on an object.

Words and Phrases That Describe Position and Direction

Position	Direction
• Above/below	• Forward/backward
• Over/under	• Toward/away from
• On top of/underneath	• Up/down
• Inside/outside	• Right/left
• In front of/behind	• North/south/east/west
• To the right/left of	• Around/through
• Next to/far away from	• Into/out of

Measuring Motion

How fast do you go when you ride your bicycle? Can you make your bicycle go as fast as a car? How fast does a car go? **Speed** is a measure of how far an object moves in a certain amount of time. Speed is how fast an object moves.

Suppose you are at the zoo watching cheetahs. You see a cheetah running, and you want to know its speed. To tell how fast an object moves, you need to know two things. First, you need to know the distance. *Distance* is how far the object moves. Second, you need to know how much time it took to move that distance. You can measure distance using a meterstick or tape measure. You can measure time using a stopwatch.

You start your stopwatch as the cheetah passes one part of the fence. You stop your watch when the cheetah passes a pole next to the fence. Then you measure how far apart the poles are. You record your data in a table.

Cheetah Speed

Distance (in meters)	Time (in seconds)
100	10
200	20
300	30

The cheetah ran 300 meters in 30 seconds. To find the cheetah's speed, you divide the distance by the time.

Speed = distance ÷ time
Speed = 300 m ÷ 30 s
Speed = 10 m/s

The unit m/s means "meters per second." The cheetah ran at a speed of 10 meters per second. This means that the cheetah ran 10 meters during every second it was running.

You can show motion on a graph. You learned about different kinds of graphs in Lesson 7. The graph below shows how time and position are related. You can see the time the cheetah took to run from one position to another.

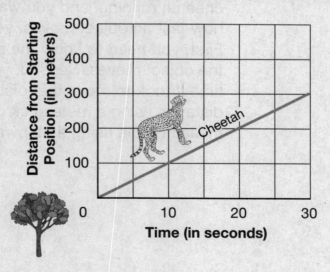

Reading a graph is not hard. The line tells you what is happening. Follow where it hits on the grid. The grid lines lead you to the numbers you need. The chart shows that the cheetah took 10 seconds to run 100 meters. If the cheetah ran faster, the line would go up at a steeper angle.

DISCUSSION QUESTION

You kick a soccer ball, and the ball rolls toward your friend. Then your friend kicks the ball toward you. Why does the ball change position? Why does it change direction?

LESSON REVIEW

1. Wind pushes a toy sailboat across a pond. The boat moves in a straight path. What is changing?

    A. the boat's direction
    B. the boat's size
    C. the boat's force
    D. the boat's position

2. Suppose you push on an object. Your push CANNOT

    A. stop the object's motion.
    B. start the object moving.
    C. change the object's mass.
    D. change the object's direction.

3. What is speed?

    A. how much an object's position changes
    B. how far an object moves
    C. how much force is needed to make an object move
    D. how fast an object moves

Lesson 32 How Forces Affect Motion

M2.1b; PS5.1b, c, d; PS5.2a

Getting the Idea

Key Words
force
gravity
mass
friction

You have learned that a **force** is a push or a pull. If someone pushes a book, it slides across the desk. You can see the person use force to push the book. But what happens when the book reaches the edge of the desk? You do not see anyone use a downward force on the book, but it falls to the floor. The book falls because of gravity. Gravity is a force you cannot see. You can only see how it affects different objects.

Force and Motion

If an object is not moving, a force can make it start moving. The object moves in the direction in which the force acts. For example, look at the soccer ball below. At first, it is not moving. If you kick the ball, you exert a force on it. The ball moves in the same direction as your kick.

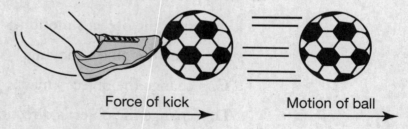

Force of kick   Motion of ball

If an object is already moving, a force in the same direction increases how fast it is moving. A force in the opposite direction decreases how fast it is moving or even makes it stop moving. When the total force pushing or pulling on an object changes, the motion of the object also changes.

Sometimes forces do not make things move. Look at the picture below.

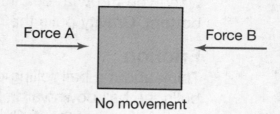

No movement

One force pushes to the right. An equal force pushes to the left. The forces are the same, so the object does not move. The forces are balanced. Now look at this picture.

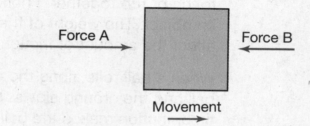

Movement

The object moves because the forces are not balanced. It moves in the same direction as the stronger force. Unbalanced forces can change an object's direction. Think about a ball rolling toward you. If you kick the ball, the forces on the ball are not balanced. The ball changes direction.

Gravity

If you hold a ball and let go of it, it falls to the ground. The force of gravity acts on the ball. **Gravity** is a force that pulls objects toward each other. The force of gravity acts on all objects. On Earth, all objects are pulled toward the planet's center. Without gravity, everything on Earth would float away.

The force of gravity depends on the mass of the object. Recall that **mass** is the amount of matter in an object. You feel the force of gravity as weight. Think about a bowling ball and a tennis ball. The bowling ball has more mass than the tennis ball. The force of gravity is stronger on the bowling ball than it is on the tennis ball. So, the bowling ball weighs more than the tennis ball.

The force of gravity can affect objects through all different kinds of matter—solids, liquids, and gases. For example, when you throw a rock into a pond, the rock sinks to the bottom. Gravity pulls the rock downward.

Friction

Think about a ball rolling down a hill. The force of gravity pulls the ball downward. The ball will continue to roll until another force stops it. Suppose the ball runs into a wall. The wall stops the ball from rolling.

Another force that can stop the ball from rolling is friction. **Friction** is a force that acts between two objects that touch or rub together. Friction tends to slow the motion of an object. The weight of the object and the type of surface affect the amount of friction.

When a ball rolls along the ground, friction between the ball and the ground slows the ball's motion. After some time, friction makes the ball stop rolling. If you push a book across a desk, friction pushes back the other way.

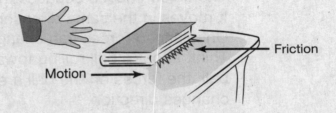

DISCUSSION QUESTION

You kick a volleyball toward your friend. Then your friend kicks the ball back toward you with the same force. Are the forces on the ball balanced? Why does the ball change direction?

Lesson 32: **How Forces Affect Motion**

LESSON REVIEW

Test Tips...

Some questions may include a diagram. Look at the diagram carefully. Figure out what the diagram shows. Find out how it connects to the question before you answer.

Use the illustration below to answer question 1.

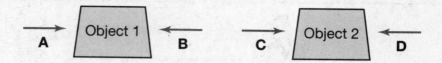

1. Which two forces should be stronger if you want the objects to move closer together?

   A. A and B
   B. C and D
   C. A and D
   D. B and C

2. Two students are about the same size. They pull on opposite ends of a long rope. The rope moves toward Student A. Which statement is TRUE?

   A. Both students pull with the same force.
   B. The force applied by Student B is stronger than the force applied by Student A.
   C. Student A applies a stronger force on the rope than Student B.
   D. The forces on the rope must be balanced.

3. Friction is a force that

   A. pulls objects toward the center of Earth.
   B. acts when two objects rub against each other.
   C. pulls two objects apart.
   D. makes an object moves faster.

Lesson

33 Magnets

PS5.1e; PS5.2a, b

Getting the Idea

Key Words
magnet
attract
pole
repel
magnetic field

Playing with magnets is fun because they stick to some things. You have probably tried to pick up many different objects with a magnet. You might have noticed that you can pick up things such as staples and paper clips.

What Magnets Attract

You feel a pull if you bring a magnet near certain metal things. A **magnet** is an object that can push or pull on some metals without touching them. Most magnets are made of iron or steel. Some rocks are natural magnets.

You learned in Lesson 22 that magnets attract only some objects. Magnets attract things made with metals such as iron or steel. To **attract** means to pull closer. If you bring a magnet near some steel paper clips, the magnet attracts the paper clips.

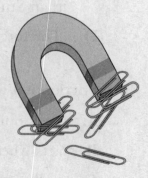

Magnets do not attract all metals. A magnet will not attract a soda can made of aluminum. A magnet will not attract a penny. Pennies are made of zinc and other metals that a magnet will not attract. Magnets will not attract things not made of metal. For example, a magnet will not attract a piece of paper, a rubber ball, or a wooden pencil.

Lesson 33: Magnets

Did You Know?

Some farmers feed magnets to their cows! A magnet can stay in a cow's stomach through its whole life. The magnet attracts small pieces of metal such as nails and staples that get mixed up in the cow's hay. The magnet holds on to the little pieces and stops them from hurting the cow.

Magnets are very useful. Maybe you have used a magnet to attach a paper to your family's refrigerator. This works because the refrigerator door contains iron. You might have noticed other ways magnets are used around the home. For example, magnets sometimes hold cabinet doors closed. Magnets may also hold curtains in place.

Magnetic Poles

A **pole** is the end of a magnet. This is where its magnetic force is strongest. All magnets have two poles. The two ends of a magnet are called its north pole and south pole. You may see these marked as N for north pole and S for south pole. Two types of magnets are a bar magnet and a horseshoe magnet. Both have a north pole and a south pole.

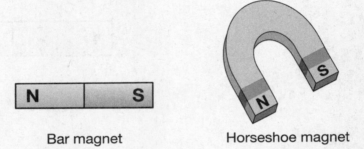

Bar magnet   Horseshoe magnet

You can feel a push or pull if you bring two magnets close to each other. If a north pole and a south pole point toward each other, the magnets attract each other. Magnets can also repel each other. To **repel** means to push away. If the north poles point toward each other, the magnets repel each other. The magnets also repel each other if the south poles point toward each other.

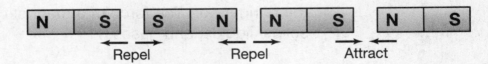

Did You Know?

The needle of a compass is a magnet. Earth has a magnetic field around it, like the field around a bar magnet. The compass needle lines up with the field and always points to the north.

Weak or Strong?

The force of a magnet is stronger when an object is closer to the magnet. The force gets weaker the farther an object is from the magnet. The area around a magnet where its force can act is called its **magnetic field**.

A magnetic field is strongest at the magnet's poles. Suppose you sprinkled some iron filings around a bar magnet. Most of the filings would be attracted to the ends of the magnet. The bar magnet is strongest at its ends. It is weakest in the middle. The drawing below shows the magnetic field of a bar magnet.

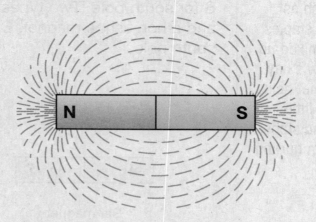

With a horseshoe magnet, iron filings are also attracted most to the ends of the magnet. This is where the magnetic field is strongest. The curved part of the horseshoe magnet has the weakest pull.

The force of a magnet can affect objects even when there is matter between the magnet and the object it attracts. You know that a magnet sticks to a refrigerator, even when there is a piece of paper between the magnet and the refrigerator. The force of a magnet can act through solids, liquids, and gases.

DISCUSSION QUESTION

Try attracting different objects with a magnet. Notice the objects the magnet attracts and the objects it does not. Classify different substances around you based on magnetism. What does this show you about the different objects?

Lesson 33: **Magnets**

LESSON REVIEW

1. You test four objects against each end of a bar magnet. One of the objects is attracted to both ends of the magnet. Which of the four objects would that be?

   A. a wooden ruler

   B. a copper penny

   C. a steel nail

   D. another bar magnet

2. The picture below shows a horseshoe magnet.

   The magnet is picking up some of the pieces in a mixture. Those pieces are MOST LIKELY made of

   A. rubber.

   B. iron.

   C. sand.

   D. aluminum.

3. What do all magnets have in common?

   A. They are all made of steel.

   B. They have a north and a south pole.

   C. They stick to wood.

   D. They only attract rubber objects.

Lesson 34: Simple Machines

PS5.1d, f

Getting the Idea

Simple Machines

Every time you pick up a book, climb the stairs, or open a door, you are doing work. In science, you do **work** when you use a force to move an object. Recall that a force is a push or a pull. Simple machines make it easier to do work.

What Is a Simple Machine?

A **simple machine** is a tool that makes work easier. It does not do work for you. It changes the way your force moves something. A simple machine has few or no moving parts. The inclined plane, wedge, screw, lever, pulley, and wheel and axle are the six kinds of simple machines.

Kinds of Simple Machines

An **inclined plane** is a slanted surface that connects a lower level with a higher level. A wheelchair ramp is an inclined plane. The ramp in the picture below is also an inclined plane.

Key Words
- work
- simple machine
- inclined plane
- wedge
- lever
- fulcrum
- screw
- pulley
- wheel and axle
- compound machine

The girl would not be able to lift the box straight up from the ground. It takes less force to push the box up the ramp. The girl has to move the box over a longer distance, but she uses less force. The work is easier because it is spread out over a longer distance.

Lesson 34: **Simple Machines**

When you use a simple machine, you use mechanical energy to move something. Not all of this energy moves the object. Recall from Lesson 32 that friction is a force that tends to slow down motion. When the girl uses the ramp, there is friction between the box and the ramp. Some energy is lost working against friction. Most machines lose some energy because of friction. This energy changes to heat.

A **wedge** is made of two inclined planes placed back to back. You push a wedge into an object. A wedge changes the direction of the force so that it pushes sideways in two directions. Wedges are often used to split logs in half. A knife is also a wedge.

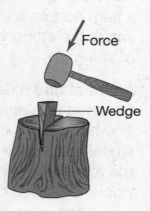

A **lever** is a bar that rests on a support called a **fulcrum**. Many common objects are levers. A seesaw is a lever. So is a bottle opener. A lever changes the direction in which you use a force. Pushing down on one end of the lever lifts an object at the other end. The picture below shows a stick used as a lever. The fulcrum is the small rock under the stick. Pushing down on the long end of the stick will lift the large rock at the other end.

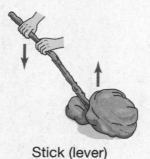

Stick (lever)

A **screw** is an inclined plane wrapped around a center pole. When you turn a screw, the work is spread over the whole distance of the inclined plane. You use less force, so the work is easier.

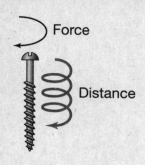

A **pulley** is a wheel with a rope or chain around it. The rope fits in a groove in the wheel. Pulling down on one end of the rope lifts an object fastened to the other end. A flagpole has a pulley for raising and lowering a flag. A simple pulley makes work easier by letting you pull down instead of lifting up.

Pulley

A **wheel and axle** is a wheel attached to a smaller round bar. The bar is the axle. Turning the wheel makes the axle turn. Turning the axle makes the wheel turn. The doorknob pictured at the right is a wheel and axle. It takes less force to turn the wheel than it would to turn the axle by itself. The wheel is bigger than the axle, so the work is spread over a larger distance.

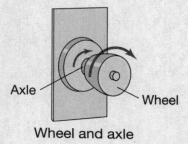

Wheel and axle

Cars, scooters, and in-line skates all have wheels and axles. A screwdriver is also a wheel and axle. The handle is the wheel, and the thin shaft is the axle.

Lesson 34: Simple Machines

Compound Machines

Simple machines can be combined to help make work easier. A **compound machine** is made up of two or more simple machines.

A bicycle is a compound machine. The brake handles are levers. The wheels and pedals make up wheel-and-axle systems. Think about the pedals. When you move the pedals, you turn an axle. It turns a larger wheel. Then the chain wrapped around the wheel moves. The chain is a pulley. Pulling on one part of the chain causes another part to move. In this case, the pulley turns the axle of the back wheel. This makes the bicycle move.

A pair of scissors is also a compound machine. The two halves of the scissors are levers. The place where they are held together is the fulcrum. The sharp blades of the scissors are wedges.

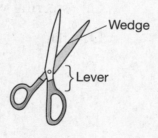

DISCUSSION QUESTION

A screwdriver is a wheel and axle. But you can also use a screwdriver to pry up the lid on a can of paint. In that case, what simple machine is the screwdriver? Explain.

Test Tips...

Some questions ask you to find more than one thing. Make sure you find everything that the question asks for. Some answer choices may list only part of the answer. If you are not careful, you may miss the correct answer.

LESSON REVIEW

1. What is TRUE about all simple machines?

    A. They all lift things.

    B. They all have moving parts.

    C. They all make work easier.

    D. They all have just one part.

2. A worker needs to lift a heavy concrete block from the ground to the second floor of a building. What would be the BEST simple machine for the worker to use?

    A. wheel and axle

    B. lever

    C. pulley

    D. screw

3. Which one is NOT an example of a simple machine?

    A. bottle opener

    B. wheelchair ramp

    C. steering wheel

    D. flag

Chapter 8 Review

1. Look at the diagram below.

   Which statement *best* describes the position of the saucer?

   A The saucer is above the cup.
   B The saucer is underneath the cup.
   C The saucer matches the cup.
   D The saucer is inside the cup.

2. What happens when you pull on a door?

   A It changes position.
   B You change the amount of force the door has.
   C Its force of gravity becomes greater.
   D Its force of friction becomes greater.

3. You stand in front of your desk. What is the position of your desk?

   A The desk is behind you.
   B The desk is next to you.
   C The desk is underneath you.
   D The desk is above you.

4 The diagram below shows a baseball bat hitting a ball.

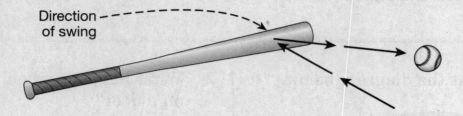

What happens to the ball after the bat hits it?

A The ball changes direction.
B The ball loses its mass.
C The ball weighs more.
D The ball stops moving.

5 The force of gravity can act on objects

A only through gases
B through solids and liquids
C through liquids and gases
D through solids, liquids, and gases

6 Which of these statements about magnets is *true*?

A Two south poles attract each other.
B Two north poles attract each other.
C A north pole and a south pole attract each other.
D A north pole and a south pole repel each other.

7 Where is the force of a magnet greatest?

A at the north pole   C in the middle
B at both poles   D at the south pole

8 Look at the diagram below.

What kind of simple machine are the students using?

A lever
B pulley
C inclined plane
D wheel and axle

9 The diagram below shows a man pushing a box up a ramp.

What is true about the inclined plane shown in this drawing?

A It stops work from being done.
B It makes the man push harder.
C It makes it easier to do work.
D It makes the box weigh less.

10 When you use a machine, some of your mechanical energy does not do work. Friction causes some mechanical energy to change to

A heat
B sound
C electrical energy
D light

11 When you squeeze the handle brakes to stop a bicycle, rubber pads press against the tire. What force causes the bicycle to stop moving?

　A　a wheel and axle
　B　gravity
　C　friction
　D　weight

12 The south pole of one magnet and the north pole of another magnet attract each other. If you move the magnets farther apart, what happens to the magnetic force between the magnets?

_____

_____

13 Why does an object fall when you let go of it?

_____

_____

CHAPTER 9: Characteristics of Living Things

Lesson 35 What It Means to Be Alive

LE1.1a, b, c, d; LE1.2a; LE4.2a, b; LE5.1a; LE5.2a

Getting the Idea

When you see a bird fly across the sky, you know it is alive. A pile of rocks is not alive. But how do you know? Living things do many things that nonliving things cannot do. Something that is **alive** can move, grow, and respond to its environment. In this lesson, you will learn how to tell what is alive and what is not.

Key Words
alive
nutrient
waste
reproduce
offspring

What Makes Something Alive?

Living things can move. Flying is one way that a bird can move. Birds also walk, and some can swim. Now think of other living things. Worms, fish, crickets, and puppies are all able to move. Even plants can move in certain ways. Sunflowers move to face the sun. New leaves and flowers move when they unfold. Every living thing has at least some parts that can move. If something is able to move on its own, you can tell that it is alive.

Have you ever seen a baby bird that has just hatched? Baby birds are small compared to their parents. Most do not even have feathers yet. As a baby bird gets older, it gets bigger. It grows feathers. Its mouth changes shape. Its wings get longer and stronger.

Lesson 35: What It Means to Be Alive

When you were born, you were much smaller, too. You most likely had no teeth, and you probably did not have much hair. You have grown and changed in many ways since then. Like you, things that are alive grow and develop as they get older.

You may have watched birds eating seeds or insects. The seeds and insects are food for the birds. In order to grow, living things must take in water and nutrients. A **nutrient** is a substance that gives a living thing energy and materials for growth. All animals get the nutrients they need by eating food. They also breathe air, which helps them use their food. Plants do not eat food. They take in sunlight, air, and water to make the food they need to grow. Part of the food plants make may become food for animals.

Birds leave droppings on the ground below them. This is another way to tell they are alive. A living thing does not use everything it takes in. Some of the food, air, and water are left over. The unused material is called **waste**. All living things get rid of waste.

When a bird sees a cat, the bird may fly away. You get hungry when you smell something good to eat. Each of these actions is called a response. A response is what a living thing does after it learns something about its surroundings. Living things often respond right away. If an animal learns danger is near, it may run away. If it smells food, it may feel more hungry.

It might surprise you to learn that plants respond, too. A plant growing toward a light is responding to the light. So is a flower opening in spring. A sunflower responds to the sun by turning to face it.

Birds hatch from eggs. A mother bird lays the eggs, and the mother and father birds may bring back food so the baby birds can grow. Like birds, all living things **reproduce**, or make more of their own kind. Living things have young called **offspring**. The offspring are the same kind of living thing as their parents. Sparrows make more sparrows. People make more people. Oak trees make more oak trees.

Nonliving Things

At some time in its life, every living thing stops living, or dies. It stops moving, growing, and responding. It stops taking in nutrients and making wastes. It is no longer alive. But it is not the same as a nonliving thing.

A nonliving thing was never alive. Nonliving things found in nature include rocks, soil, water, and air. Many nonliving things, such as cars and computers, are made by people.

Nonliving things do not move on their own. They do not respond to their surroundings. They do not take in nutrients. And they do not make more of their own kind.

Sometimes, nonliving things can seem like living things. At times, a rock may roll or fall. The rock moves. But is it alive? A rock only moves if it is pushed or pulled. Now think about a cloud. A cloud may move and get bigger. But a cloud does not make more clouds. Clouds do not eat or make food. New clouds form from water in the air. What about soil? More soil is made when rocks break down. The soil cannot make more of itself.

Rocks   Clouds   Soil   Water

Rocks, clouds, rivers, and soil are not alive. To be alive, something must be able to do all the things plants and animals can do.

Lesson 35: What It Means to Be Alive

DISCUSSION QUESTION
How is the way a living thing moves different from the way a cloud moves?

_____

LESSON REVIEW

1. Which of these is TRUE of something that is alive?
    A. It is a certain size.
    B. It makes noise.
    C. It is a certain color.
    D. It can respond to things.

2. Because animals take in food, water, and air, they must also
    A. have young.
    B. move quickly.
    C. get rid of wastes.
    D. take in sunlight.

3. A nonliving thing does NOT
    A. eat.
    B. fall.
    C. break.
    D. move.

4. Which two are examples of living things?
    A. birds and rocks
    B. humans and clouds
    C. humans and birds
    D. clouds and rocks

Lesson 36
Inherited, Acquired, and Learned Traits

LE2.1a, b; LE2.2a, b; LE5.2f; LE6.1e

Getting the Idea

Key Words
trait
inherited trait
species
offspring
acquired trait
behavior
instinct
migrate
hibernate

How would you describe yourself? Maybe your eyes are brown. You may like to skate. Others in your family may have brown eyes, too. But others in your family may not skate. Your eye color came from your parents. However, you learned how to skate. Humans and other living things have many different characteristics, or **traits**. Some traits you got from your parents before you were born. Other traits you have picked up during your lifetime.

Inherited Traits

An **inherited trait** is a characteristic that a living thing gets from its parents before birth. People in the same family tend to look similar. The color of your eyes, hair, and skin are traits that you inherited from your parents. These traits also include curly or straight hair, dimples and freckles, and a tendency to be short or tall.

A tiger's stripes, a leaf's shape, and the color of a person's eyes are all inherited traits.

All living things have inherited traits. Plants and animals look like their parents and like other members of their species. A **species** is a kind of living thing. When plants and animals reproduce, they pass on traits to their babies, or **offspring**. A horse's hair color is inherited from its parents. Two black cats are likely to have black kittens. Seeds from a pink rose bush are likely to grow into more pink rose bushes. Puppies from the same litter might look a lot like their parents.

Lesson 36: **Inherited, Acquired, and Learned Traits**

Acquired Traits

An **acquired trait** is a trait that a living thing picks up during its life. Living things do not get acquired traits from their parents before they are born. These traits are not passed on to offspring. A living thing gets an acquired trait by interacting with its environment.

An acquired trait can be a change to the body, such as getting a scar. A cat may hurt one of its paws. But when the cat reproduces, the kittens have normal paws. The acquired trait is not passed on to its offspring.

Inherited Behaviors

Some traits are **behaviors**, or ways of acting. Animals have behaviors that help them survive in their environments. Certain behaviors can be inherited. Inherited behaviors are called **instincts**. Many animals are born with instincts that help them survive. A bird knows how to build a nest by instinct. A spider spins a web by instinct, too.

Different animals in different environments show different inherited behaviors. As winter approaches, many animals have a harder time finding food. Geese, caribou, monarch butterflies, whales, and other animals **migrate**. They move from one place to another in a pattern. They go to warmer areas or follow food sources. In spring, they return to their homes.

As the weather gets colder, chipmunks, bats, mice, and other animals eat a lot and then **hibernate**. They go into a sleep-like state during winter and live off the fat stored in their bodies until spring.

Fish, amphibians, and reptiles often become dormant, or inactive, during winter. When dormant, an animal does not move much and needs little food or air. In a pond, dormant fish and frogs rest at the bottom of the water. No one taught these animals how to survive the winter. They know what to do by instinct.

Many animals that live in hot, dry deserts sleep during the day. They are active at night, when the air is cooler. This is an inherited behavior that helps them survive.

Learned Behaviors

Are you better at using a computer than the adults in your family? If so, you learned how to do this. You did not inherit this behavior from your parents before you were born. After an animal is born, it begins to learn.

An animal may learn skills from its parents. A mother bear shows her cubs how to find food. A human baby learns to speak by listening to and copying others. These are behaviors that an animal learns over time.

Things you learn to do are acquired traits. Parents cannot pass acquired traits to their offspring when they reproduce. The offspring have to learn to do these things, too.

In the wild, young animals need to learn many skills to survive. Sometimes they learn these skills by watching other animals. Sometimes they learn these behaviors by accident. A deer that is harmed by a hunter might learn to stay away from people.

Learned behaviors are different from inherited traits because they can change over time. You were born with two legs. That is an inherited trait. However, you can become a good hockey player by practicing and figuring out the rules. Hockey skills are learned behaviors.

DISCUSSION QUESTION

Think of some of your behaviors. How can you tell whether a behavior was inherited or learned?

LESSON REVIEW

1. Which of these is an inherited trait in humans?
    A. having a scar
    B. having long fingers
    C. having short hair
    D. speaking English

Lesson 36: Inherited, Acquired, and Learned Traits

2. Which of these is a learned behavior?

   A. having long eyelashes

   B. migrating

   C. hibernating

   D. speaking English

3. Which of these can you change?

   A. your instincts

   B. your inherited traits

   C. your learned behaviors

   D. your inherited behaviors

4. Your neighbors are very good cooks. Why are their young children likely to become good cooks when they are older?

   A. They were born knowing how to cook.

   B. They will learn to cook by watching their parents.

   C. They already have their cooking skills.

   D. Cooking is an inherited trait.

5. The picture below shows a flock of geese.

   What are the geese doing to help them survive?

   A. hibernating   C. migrating

   B. sleeping   D. learning instincts

Lesson 37: Plant Structures and Functions

LE3.1b, c; LE5.1b; LE5.2a; CT1

Getting the Idea

A tree, such as an apple tree, is a living plant with several main parts. So is a tomato plant. Those two plants look very different, but they have the same basic needs. They also have similar parts to meet those needs. The parts of a plant work together to keep the plant alive and healthy.

Key Words
- root
- stem
- leaf
- photosynthesis
- reproduce
- flower
- seed
- fruit
- adaptation
- germinate

Parts of a Plant

Most plants have a body made up of roots, stems, and leaves. The picture below shows the roots, stems, and leaves of a tomato plant. The plant needs all three parts in order to survive, or stay alive.

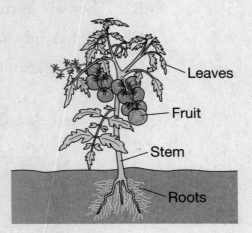

A **root** is a plant part that takes in water and nutrients from the soil. A plant's roots usually grow underground. They hold the plant in the soil. The roots of some plants also store some of the food that the leaves make.

212 Duplicating any part of this book is prohibited by law.

Lesson 37: Plant Structures and Functions

Did You Know?

Some rainforest plants have roots that never touch the ground. The roots take up water from the damp air.

Most plants' roots branch out as they grow downward. The branching roots of a plant form a root system. Root systems can be very small or very large. They can be deep or shallow. The roots of tall trees often spread out many meters around the trees.

Stems

A plant's **stem** moves water and nutrients from the roots to the leaves and other parts of the plant. The stem also supports the leaves, flowers, and fruit. Like roots, stems sometimes store food. In some kinds of plants, stems make food, too.

Some plant stems are green and flexible, while others are brown and stiff. Tomato plants and sunflowers have green stems. Trees and shrubs have brown stems made up of wood and bark. But all stems do the same kinds of jobs to help plants stay alive.

Leaves

A **leaf** is a plant part that captures the sun's energy and makes food for the plant. A green material in the leaf and stem takes in sunlight. Tiny parts of the leaf use light energy to turn carbon dioxide and water into food. The process by which plants use the energy of sunlight to make food is called **photosynthesis**. Many leaves are broad and flat, but leaves come in all shapes and sizes. The needles of a pine tree are long, thin leaves. Like other leaves, the needles make food for the tree.

Flowers and Seeds

Plants also have parts that help them **reproduce**, or make more of their own kind. In most plants, these parts are flowers and seeds. A **flower** is a plant part in which seeds form. A **seed** contains a tiny new plant. The seed protects the new plant and stores food to help it start growing.

Fruits form around the seeds. You have seen seeds inside apples and cherries. You might be surprised at some of the things that are fruits. Tomatoes are fruits. So are cucumbers. The fluffy parts that you blow from a dandelion are fruits, too.

Plant Adaptations

Each of a plant's parts is an adaptation. An **adaptation** is a body part or behavior that helps a living thing stay alive. Plants that live in different kinds of places have the same basic parts, but they have different adaptations. Their parts look different. For example, plants that live in shady places tend to have large, wide leaves. These leaves take in as much sunlight as possible. Plants that live in sunny places tend to have smaller leaves. They do not need large leaves to get enough sunlight.

Deserts are places with very little rainfall. When it does rain, the rainwater dries up before it can soak down very far. So the roots of desert plants tend to spread out near the surface. They take in water quickly from a big area. Then the water is stored in their thick stems. Their roots and thick stems are adaptations for getting and holding water.

Desert plants have small leaves. Plants lose water through their leaves. Small leaves keep a plant from losing too much water. The cactus shown here has such small leaves that you would need a microscope to see them. They are too small to make enough food for the plant. Instead, its thick green stem captures sunlight and makes food. The cactus in the picture has another adaptation—its sharp spines. They keep animals from eating the plant.

Cactus Adaptation

Lesson 37: Plant Structures and Functions

Plants have many adaptations for reproduction. Seeds need water and soil to germinate. To **germinate** means to sprout. Recall that fruits form around seeds. You have eaten apples. You threw away the core when you were done. An animal such as a squirrel does the same thing. If the squirrel drops the core in a place that is good for growth, the seeds germinate. Sweet, juicy fruit is an adaptation that helps move the seeds to a good place to grow.

Not all fruits are sweet and juicy. Have you picked burrs off clothes after walking through the woods or a field? The rough covering of the burr is a fruit. The burr sticks to a person or an animal. The burr falls off in a new place, where the seed may sprout. The fruit of a maple seed is shaped like a wing. When the wind blows, the seeds fly in the wind. This is an adaptation that helps the seeds move to a good place to grow.

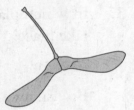

DISCUSSION QUESTION

All living things need water to survive. How do roots and stems help plants get the water they need? How do leaves use water?

LESSON REVIEW

1. What does a plant root do?
    - A. turns carbon dioxide and water into food
    - B. takes in water and nutrients from the soil
    - C. supports the leaves
    - D. makes seeds

2. Which plant parts help plants reproduce?
    - A. roots and stems
    - B. stems and leaves
    - C. flowers and seeds
    - D. needles and roots

3. In which plant part does photosynthesis usually take place?
    - A. root
    - B. stem
    - C. leaf
    - D. flower

4. Which plant parts show adaptations?
    - A. mainly roots
    - B. mainly leaves
    - C. mainly stems
    - D. all plant parts

Lesson 38: Life Cycles of Plants

LE4.1a, b, c, d

Getting the Idea

All living things begin life, grow and change, reproduce, and later die. All the stages of a living thing from the beginning of life to death make up a **life cycle**. Every plant goes through these basic stages. But different kinds of plants have different life cycles. How plants start off depends on how the parent plants reproduce. Plants also grow and change in different ways.

Key Words
- life cycle
- germinate
- pollen
- life span
- cone
- spore
- bud
- runner

Growing from Seeds

You learned about seeds in Lesson 37. Most plants reproduce by making seeds. All flowering plants reproduce in this way. The diagram below shows the life cycle of a flowering plant.

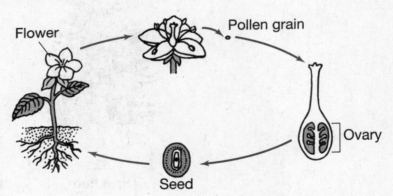

The starting part of a new plant is inside the seed. The seed also contains stored food. This food helps the new plant grow when the seed germinates. Recall that to **germinate** means to sprout. When a new plant starts growing from a seed, a new life cycle begins. The new plant grows parts such as roots, a stem, and leaves. It gets bigger and taller. When the plant is an adult, it develops flowers. Then the plant is ready to reproduce.

Flowers produce a powdery material called **pollen**. Bees and other insects spread pollen from flower to flower. Wind may also spread pollen. When flowers receive pollen, seeds form in parts of the flowers called ovaries. A fruit, which you learned about in Lesson 37, is the ripened ovary of a flowering plant. New plants may grow from some of the seeds, and the cycle begins again.

After a time, a plant dies. Each kind of living thing has its own **life span**, the amount of time it lives. Some plants complete their life cycles in only one year. Others, such as trees, live for many years.

Pine trees and some other plants produce **cones** instead of flowers. Male cones contain pollen. Animals and the wind carry pollen from male cones to female cones. Then seeds form inside the female cones. Some of the seeds drop to the ground and sprout, or germinate. New life cycles begin.

Pine tree   Cone and needles

Growing from Spores
Some plants do not produce flowers, pollen, seeds, or cones. Instead, they produce spores. A **spore** is a plant cell that can grow into a new plant. Spores drop off the plant into the soil.

Most spores need very moist soil in order to sprout. Some spores have a protective coating. They can lie in soil until there is enough moisture for them to sprout. Then a new life cycle begins.

Mosses and ferns are kinds of plants that use spores to reproduce. The picture on the right below shows a fern frond, or leaf, with spores.

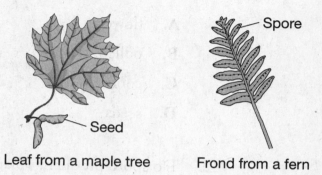

Leaf from a maple tree   Frond from a fern

Growing from the Parent Plant

Some kinds of plants can grow either from seeds or from part of the parent plant. For example, a new potato plant can grow from seeds or from an "eye" of a potato. The "eye" of a potato is a **bud**, a plant part that can grow into a new stem, leaf, or flower. Many plants such as strawberry plants and some kinds of grass send out runners. A **runner** is a kind of stem that grows sideways on the soil surface. Buds form along the runner and grow into new plants.

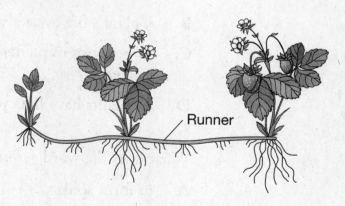

DISCUSSION QUESTION

People who are allergic to pollen can have allergy problems in the spring, summer, or fall. What do you think that tells you about the life cycles of plants in the area?

LESSON REVIEW

1. Pine trees do NOT produce
    A. flowers.
    B. pollen.
    C. cones.
    D. seeds.

2. Do spore-producing plants need bees and other insects to help them reproduce?
    A. Yes, because birds and bees help spread their pollen.
    B. No, because they have no pollen to spread.
    C. Yes, because they have no flowers.
    D. No, because they have no cones.

3. Which statement is TRUE?
    A. Some kinds of plants do not have life cycles.
    B. A plant's life cycle always starts with a seed.
    C. Some kinds of plants do not reproduce during their life cycles.
    D. All plants have life cycles.

4. What does the word *germinate* mean?
    A. to form seeds
    B. to produce germs
    C. to form spores
    D. to sprout

Lesson 39 Animal Structures and Functions

LE3.1a, c; LE5.1b; LE5.2b, c, d, e; CT1

Getting the Idea

Key Words
beak
sense organ
brain
camouflage
adaptation

An eagle's wings are for flying. A goldfish has fins and a tail for swimming. Your legs are for walking, running, and jumping. Every animal has body parts that help the animal survive, or stay alive.

Moving

Animals have body parts that let them move from place to place. Animals with legs walk, run, jump, and climb. Monkeys use their arms to swing through trees. Most birds and many insects have wings that allow them to fly. Fish move their tails and fins to swim. Snakes use muscles in their long bodies to move across the ground. Being able to move helps animals find shelter and food. It also helps animals escape from animals that hunt them for food.

Eating and Drinking

Animals have body parts that let them take in food and water. Meat eaters have pointed teeth for biting and tearing. Plant eaters have flat teeth for grinding. Humans have both pointed teeth and flat teeth, for eating both meat and plants.

The hard, hornlike part of a bird's mouth and jaws is called its **beak**. Birds have different kinds of beaks for eating different kinds of foods. A heron has a long beak, which it uses to spear fish. A cardinal has a short, heavy beak, which it uses to crack open seeds.

Duplicating any part of this book is prohibited by law.

Did You Know?

Rattlesnakes have sense organs called *pits*. They look like small holes near the snakes' eyes. The pits let rattlesnakes sense the body heat of other animals. This helps the snakes hunt animals for food.

Caterpillars have jaws for biting plants. But adult butterflies do not have jaws. A butterfly has a mouth that is shaped like a drinking straw. The butterfly uses it to suck a sweet liquid called nectar from flowers.

Senses and Communication

Animals have body parts for learning about their surroundings. These body parts are called **sense organs**. Your sense organs are your eyes, ears, nose, tongue, and skin. They are the organs that help you gather information about what is around you. Many kinds of animals have sense organs similar to yours.

Each animal's sense organs are suited to the way it lives. For example, large eyes help owls see when they hunt at night. Moles live underground and do not need to see much. They have very small eyes with coverings that keep dirt out. A mole's main sense organ is its nose.

Senses are important because they can tell an animal where to find food and where an enemy is. An animal's sense organs work with its **brain**, the part of the body that learns and thinks. Senses tell the brain what is happening around the animal. The brain tells the animal how to respond. For example if a dog smells food, the dog responds by making saliva. If a dog is hot, it responds by panting. If a dog is cold, it responds by shivering.

Animals also communicate in different ways. Birds make loud sounds to warn other birds if a cat is near. Dogs growl to warn other animals to stay away. Horses move their heads a certain way to tell other horses when they are scared. Skunks spray a liquid that smells bad to scare away enemies.

Body Coverings

Body coverings also help animals stay alive. Fur helps keep bears, squirrels, and other mammals warm. Feathers keep birds warm and help them fly. Small, thin plates called scales keep a snake's body from drying out. A fish's scales help it glide through water. A clam's hard shell protects its soft body. A porcupine's sharp spines keep other animals from attacking it.

Claws are part of some animals' coverings. A claw is a sharp, curved nail at the end of an animal's toe. Animals use claws to defend themselves. Claws also help them catch and kill prey.

Body coloring or patterns can help animals hide when they are hunting or being hunted. Coloring that helps an animal blend in with its surroundings is called **camouflage**. Light-looking fur on polar bears and arctic hares helps them blend into the snowy places where they live. Box turtles have yellowish markings on their green shells. These colors and patterns help the turtles blend in on the ground.

Adaptations

These body parts are **adaptations** that help animals survive. Feathers are an adaptation. Feathers can help a bird fly. They can also help keep a bird warm. Pointed teeth are an adaptation. They can help an animal eat meat. They can also scare other animals. White fur is an adaptation that helps animals such as arctic hares hide in the snow. Behaviors can be adaptations, too. Recall from Lesson 36 that some animals hibernate in winter. Other animals migrate to warmer places during winter.

DISCUSSION QUESTION

What are some ways you have noticed animals communicate without using sound?

LESSON REVIEW

1. How does camouflage help some animals survive?

    A. It helps them move around.

    B. It helps them stay warm.

    C. It helps them blend in with their surroundings.

    D. It helps them sense their surroundings.

2. Which one is a sense organ?

    A. skin

    B. fin

    C. scale

    D. jaw

3. An animal has teeth that are wide and flat. The animal MOST LIKELY eats

    A. other animals.

    B. mainly plants.

    C. both animals and plants.

    D. only fish.

4. Which body part does NOT help a fish move through water?

    A. fins

    B. tail

    C. scales

    D. jaws

Lesson 40: Animal Life Cycles

LE4.1a, e, f, g

Getting the Idea

You were once a baby. You are bigger and stronger now. You are growing up. One day you will be an adult. How will you be different then? All living things change over their lifetimes. In this lesson, you will learn how animals grow and change.

Key Words
reproduce
offspring
life span
life cycle
metamorphosis
larva
pupa

Growing and Changing

Animals, plants, and all other living things begin life, grow and change, and later die. During their lifetimes, living things **reproduce**, or make more of their own kind. A mother cat has kittens. The kittens grow up to be adult cats. Those cats may have their own kittens.

When living things reproduce, the new living things are their **offspring**. You are the offspring of your parents. As you learned in Lesson 36, offspring are always the same kind of living things as their parents. Cats produce cats. Humans produce humans. Offspring usually look like their parents in many ways.

A kitten changes as it grows into an adult cat. When it is an adult, the cat may reproduce. A living thing can reproduce only when it is an adult. After many years, the cat will grow old. One day, the cat will die. Each kind of living thing has its own **life span**, the amount of time it stays alive.

All the stages from birth to death make up a **life cycle**. You learned in Lesson 38 about plant life cycles. All living things have life cycles. When an adult reproduces, a new life cycle begins.

Animals That Do Not Change Form

Some kinds of animals give birth to live young. For example, fox cubs are the live young of a fox. They have the same form, or shape, as an adult fox. They have four legs, a head, and a tail. They look a lot like their parents.

The mother feeds and protects the cubs. As they grow, they start to care for themselves. They change, but they keep the same general shape. In time, they are adult foxes. The diagram below shows the life cycle of a fox.

Other kinds of animals lay eggs. When a baby bird comes out of an egg, the baby has the same form as its parents. It has two legs and two wings, a head, and a tail. Many kinds of birds feed and care for their young. They protect them until they learn to fly and can find their own food.

Most snakes reproduce by laying eggs. When the eggs hatch, young snakes come out that have the same shape as an adult snake. Snakes do not take care of their offspring. The young snakes are able to feed and care for themselves as they grow and change.

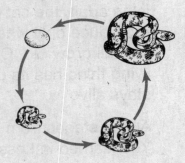

Lesson 40: Animal Life Cycles

Caterpillars have been called "eating machines." Some kinds of caterpillars double their weight every day!

Animals That Do Change Form

Some animals change form as they become adults. The change in form is called **metamorphosis**. The picture shows metamorphosis in the life cycle of a frog.

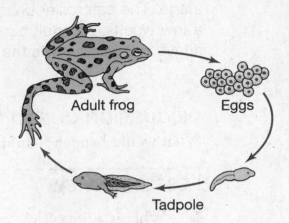

An adult frog lays eggs in water. A tadpole comes out of an egg. The tadpole lives in the water. It breathes, swims, and eats underwater like a fish. In time, it grows legs. It gets bigger and its tail disappears. Now it lives on land and in water. It breathes air. It is an adult frog.

Many insects have four stages in their life cycles—egg, larva, pupa, and adult. The picture shows metamorphosis in the life cycle of a butterfly.

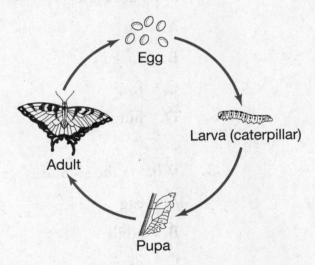

An adult butterfly lays eggs on a leaf. A caterpillar comes out of an egg. The caterpillar is the **larva** stage of a butterfly. The caterpillar eats and grows. Then it makes a hard case around itself. The insect is now in the **pupa** stage. The caterpillar changes form inside the case. After a few weeks, an adult butterfly comes out. If it is a female, it lays eggs and starts the life cycle over again.

DISCUSSION QUESTION
What would happen if animals stopped reproducing?

LESSON REVIEW

1. What is a life cycle?

   A. the way a living thing reproduces

   B. the way a living thing feeds itself

   C. one of the stages of a living thing

   D. all of the stages of life from birth to death

2. Which of these living things goes through metamorphosis in its life cycle?

   A. cat

   B. snake

   C. frog

   D. human

3. What is the second stage in the life cycle of a butterfly?

   A. egg

   B. adult

   C. pupa

   D. caterpillar

Lesson 41 Human Health

LE5.3a, b

Getting the Idea

Key Words
healthy
balanced diet
food group
exercise
sleep
habit
hygiene
germ

The clothes you wore to school last year probably don't fit you anymore. You are growing, and you will continue to grow for many more years. There are things you can do to stay healthy and help you grow up to be a healthy adult. A **healthy** person is strong, has lots energy, and doesn't get sick very often.

Balanced Diet

You know that you need food to survive. You like some foods more than other foods. Sometimes the foods that people like best are not very good for them. Too many fried foods and sugary snacks are not good for you. It is important to eat a **balanced diet**, a mix of foods that give your body the energy and nutrients it needs.

A food pyramid helps you plan a healthy balanced diet. Each main part of the pyramid is called a **food group**. Look at the food groups in the food pyramid shown on the next page.

Duplicating any part of this book is prohibited by law. 229

Some food groups make up a larger part of the pyramid than others. Grains make up the largest part of the food pyramid. Grains include bread, rice, cereal, and pasta. They give your body energy. The milk group includes cheese and yogurt. The milk group and the meat and bean group give your body nutrients you need to grow. Vegetables and fruits also give you energy and nutrients.

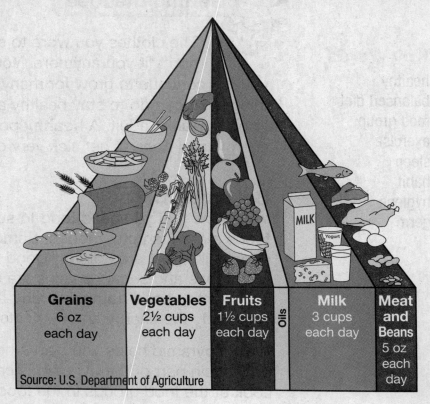

Oils are not a food group, but you do need a little oil in your diet. If you eat many fried foods, such as french fries or fried chicken, you will eat more oil than you should.

Exercise

You also need **exercise**, physical activity that keeps the body healthy. Exercise makes your muscles work. This helps them get stronger. Exercise makes your bones stronger, too. Running, walking, swimming, and playing sports are all kinds of exercise. Many everyday activities are exercise. Do you ride a bike or walk to school? That is exercise. Do you climb stairs at home or at school? That is exercise, too.

Lesson 41: Human Health

Did You Know?

Sleep is even more important than many people realize. By the time you are 75 years old, you will have spent about 23 years sleeping. That would be one-third of your entire life!

Exercise is good for your heart and the muscles that make you breathe. Do you get out of breath if you run up the stairs? If you keep using the stairs, you will make your heart and breathing muscles stronger. Pretty soon, running up a flight of stairs will be easy. It will not make you feel out of breath.

Rest

You probably find that you are tired at the end of the day. You need a good night's sleep so that you will not be tired in the morning. **Sleep** is a very important period of rest. You should get 10 or 11 hours of sleep each night. While you sleep, your muscles rest. Your body makes more cells to help you grow.

You spend a lot of time during the day thinking and making decisions. What should you wear today? Did you remember your homework? When you sleep, your brain sorts and classifies information. Sleep helps your brain. If you get enough sleep, you may even find that you remember things more easily.

When you get a good night's sleep, you wake up with a rested brain and body. People who get a good night's sleep do better in school than people who do not sleep enough. People who do not get enough sleep may fall asleep in the middle of the day. They also get angry more often. So to be happy and healthy, a good night's sleep is important.

Did You Know?

One teaspoonful of soil may contain 100 million bacteria. Some of them can make you sick. That's a very good reason to keep your hands clean!

Good Health Habits

A **habit** is something you get used to doing all the time. If you learn good health habits now, you will keep those habits all through your life. Eating a balanced diet, exercising, and getting plenty of sleep are all good health habits.

Having good **hygiene**, or keeping yourself clean, is an important habit that helps you stay healthy. **Germs** are tiny living things that cause diseases such as colds and flu. Germs are all around you. When you touch things, you pick up germs on your hands.

The best way to keep the germs out of your body is to wash your hands. Wash your hands before you eat or before you leave the bathroom. Wash your hands after you touch an animal and after you do anything that gets your hands dirty.

To get rid of germs, here is how to wash your hands:

- Run warm water over your hands.
- Use soap and rub your hands together for 15 seconds.
- Be sure to wash the backs of your hands and your fingernails.
- Rinse the soap off with warm water.
- Dry your hands on a clean towel or under an air dryer.

Good health habits also include staying away from things that are bad for you. Smoking can hurt your lungs and heart. Alcohol and drugs are harmful to a growing body. People who smoke, drink alcohol, or take drugs often find they cannot stop. It is much better to never start drinking, smoking, or taking drugs. Smart choices will help you stay healthy.

Lesson 41: **Human Health**

Test Tips...

One of the most important things you can do the night before a test is to get a good night's sleep. You will be rested and ready to do your best.

✏️ **DISCUSSION QUESTION**

Imagine that you are going on a trip to a museum the following day. How will getting a good night's sleep help you enjoy your trip?

_____

LESSON REVIEW

1. Which food group contains rice and pasta?
   A. grains
   B. vegetables
   C. fruits
   D. milk

2. Which one is NOT a good health habit?
   A. eating a balanced diet
   B. washing your hands often
   C. drinking alcohol
   D. sleeping 10 hours each night

3. What is hygiene?
   A. a kind of exercise
   B. keeping yourself clean
   C. part of the meat and bean group
   D. a kind of diet

Review

1. Which one is *not* needed by animals to stay alive?

    A water
    B air
    C sunlight
    D food

2. Which one do plants make for themselves?

    A water
    B food
    C air
    D light

3. Which one is nonliving?

    A earthworm
    B tree
    C tree stump
    D rock

4. When a living thing grows, it

    A makes food
    B gets larger
    C inherits traits
    D takes in air

5. Which one is an inherited trait in humans?

    A pierced ears
    B a scar
    C blue eyes
    D short hair

6 A living thing gets all its acquired traits from its

A mother
B father
C mother and father
D environment

7 Plants and animals pass their traits to their offspring when they

A respond
B reproduce
C take in nutrients
D migrate

8 Which plant part takes in water and nutrients from the soil?

A root
B stem
C leaf
D flower

9 Some trees grow in very windy places. Which one is an adaptation that helps a tree survive in a windy place?

A thick stems that store water
B broad leaves to get sunlight
C long roots that reach deep into the ground
D sweet fruits that attract animals

10 Which one is done by a fully grown plant?

  A germinating
  B reproducing
  C using food in seeds
  D growing a root

11 To reproduce, pine trees use pollen that is found inside

  A seeds
  B fruits
  C cones
  D leaves

12 Which kind of plant reproduces by spores?

  A fern
  B strawberry plant
  C apple tree
  D rose bush

13 The length of time a plant lives, from germination until it dies, is a plant's

  A life process
  B life span
  C life cycle
  D life growth

14 How does a plant change from a young plant to an adult plant?

   A It germinates.
   B It grows eyes.
   C It grows bigger and makes flowers or cones.
   D It uses food stored in the seed.

15 Which of these does an eagle's sharp beak help it do?

   A run from enemies
   B catch its food
   C hide from enemies
   D migrate

16 The snowshoe rabbit lives in a place that is cold and snowy in winter. In winter, the rabbit's fur turns white. The snowshoe rabbit's fur is an example of

   A learned behavior
   B mimicry
   C migration
   D camouflage

17 Animals use wings, legs, and fins to

   A eat
   B drink
   C move
   D hide

18 Sweating, panting, and making saliva are ways animals

   A respond to their surroundings
   B warn other animals of danger
   C move to a new place
   D get their food

19 Which food group contains cheese?

   A grains
   B vegetables
   C fruits
   D milk

20 A tree is shown in the diagram below.

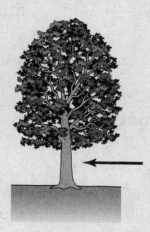

The arrow points to which part of this tree? _____

What is the function of this part?

_____

21 What are the **four** stages of a butterfly's life cycle?

(1) _____

(2) _____

(3) _____

(4) _____

CHAPTER 10
Living Things and Their Environments

Lesson 42: Producers and Consumers

S2.3a; LE6.1a, b, c, d; LE6.2a, b

Getting the Idea

You know that you get energy from the food you eat. You might be surprised to learn that almost all the food energy on Earth comes from the sun. Recall that plants make their own food. Animals cannot make their own food. All living things are divided into three main groups, depending on how they get food.

Key Words
producer
consumer
herbivore
carnivore
omnivore
decomposer
food chain

Producers

A living thing that makes its own food is called a **producer**. Plants are producers. Plants use water, carbon dioxide from the air, and the energy of sunlight to make food. You learned in Lesson 37 that this process is called *photosynthesis*. A tiny moss plant and a tall oak tree are both producers. Algae and some bacteria are producers, too. Grasses, shrubs, and trees are all producers.

Consumers

Animals, including people, cannot produce their own food. People can cook and eat food, but they cannot produce their own food from sunlight, as plants can. The word *consume* means "to eat." When you eat a food, you consume it. You and all other animals are **consumers**.

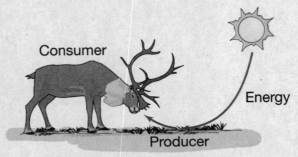

Plants absorb and trap energy in sunlight.
Animals get this energy when they eat plants.

Herbivores are consumers that eat mainly plants. In a forest, deer and rabbits are common herbivores. Horses, cows, elephants, and zebras also are herbivores.

Carnivores are consumers that eat mainly other animals. Sharks, walruses, seals, and octopuses are carnivores that live in the ocean. Lions, wolves, hawks, and eagles are carnivores that live on land. Some carnivores are scavengers. These carnivores eat animals that are already dead. Most of the time, scavengers eat leftovers from other carnivores. One example of a scavenger is a vulture.

Lion   Vulture

A lion is a carnivore. A vulture is a scavenger.

Omnivores are consumers that eat both plants and animals. Because they can eat a variety of organisms, omnivores can easily adapt to changing environments. Pigs, bears, raccoons, and humans are examples of omnivores.

Decomposers

Sooner or later, all living things die. What if they did not decay, or rot away? Dead plants and animals would form a thick layer covering Earth's surface. Instead, living things called decomposers cause dead plants and animals to decay. A **decomposer** is a living thing that gets energy by breaking down dead material from living things. Decomposers break the material down into tiny pieces that become part of soil. They add nutrients to the soil to help more plants grow. Bacteria, molds, and some mushrooms are important decomposers. Earthworms are both decomposers and consumers.

Food Chains

A **food chain** is a model that shows how living things are connected. In a food chain, energy flows from one living thing to another.

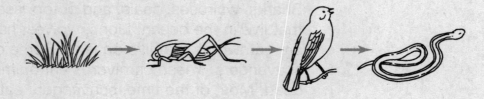

Food chains begin with a producer. In the food chain above, the grasses are producers. They use the energy of sunlight to make food, and they store some of that energy in their leaves. When the cricket eats the grass, the energy in the plant moves into the cricket's body. The cricket uses some of that energy to hop and find food. It stores the rest of the energy from the grass in its body.

When the bird eats the cricket, the energy in the cricket moves into the bird's body. The bird uses some of that energy to stay warm, find food, and fly. It stores the rest of the energy from the cricket in its body.

Then the snake eats the bird. The snake uses some of the energy from the bird and stores the rest in its body. Each living thing in this food chain gets energy from the sun or from the plant or animal it eats. However, all the energy started with the sun.

The bird eats more than just crickets. The bird may eat worms or seeds. The bird is part of more than one food chain. Living things often fit into many food chains. Many food chains can connect and overlap.

Lesson 42: Producers and Consumers

Test Tips...

After you finish taking a test, review your work. Make sure you answered every question. Make sure you understood each question. Check to see that you marked the answers you meant to mark.

DISCUSSION QUESTION

Think of a forest with trees, foxes, squirrels, birds, insects, mice, mushrooms, bacteria, and mold. Identify the role of each living thing. Is it a producer or a consumer? Which ones are decomposers? How might a food chain go that includes three of these living things?

LESSON REVIEW

1. Which of these statements is true about how living things get food?
   A. Producers eat consumers.
   B. Producers eat other producers.
   C. Consumers eat only other consumers.
   D. Consumers eat producers and other consumers.

2. Which type of animal mainly eats other animals?
   A. an omnivore
   B. a carnivore
   C. an herbivore
   D. a producer

3. After the sun, where does a food chain start?
   A. with animals that eat grass
   B. with animals that eat other animals
   C. with plants that eat animals
   D. with plants that make their own food

4. Where would a decomposer be in a food chain?
   A. at the beginning
   B. in the middle
   C. at the end
   D. Decomposers are not part of a food chain.

Lesson 43 Changes in Populations

M2.1b; LE3.2a, b; LE5.2g; LE6.1b, e, f

Getting the Idea

A living thing's surroundings make up its **environment**. The environment affects the health and growth of living things. In this lesson and the next lesson, you will learn more about the living and nonliving parts of the environment.

Key Words
environment
resource
population
community
predator
prey
variations

What Living Things Need

Living things need many **resources** in order to survive. Living things need food, water, sunlight, and air. They also need space. Most plants need soil, and animals need shelter.

Populations

A **population** is a group of organisms of the same species living in the same area. Many different species live in an area. A **community** is all the populations that live in the same area.

Populations affect each other in many ways. They may compete for resources. For example, both deer and mice eat the acorns that fall from oak trees. Dandelions and grass plants compete for space in your lawn. Members of the same species also compete for resources.

One population may feed on another. For example, hawks feed on mice. A hawk is a **predator**, an animal that hunts other animals for food. The animal the predator hunts is called the **prey**.

Lesson 43: **Changes in Populations**

Did You Know?

European settlers brought apple seeds to New York in the 1600s. They planted them and grew apple trees. Before that, there were no apples in America.

A predator population and a prey population affect each other. When the prey population is large, there is a lot of food for the predators. So the predator population gets larger. When there are more predators, they eat more prey. So, the prey population gets smaller. Then there is less food for the predators. The predator population gets smaller. When there are fewer predators, they eat less prey. Then the prey population grows larger again.

The two populations get larger and smaller in a cycle that repeats. The graph shows the sizes of two populations. The moose is the prey. The wolf is the predator. Look at 1990 and 1995. After the wolf population became very small, the moose population got much larger.

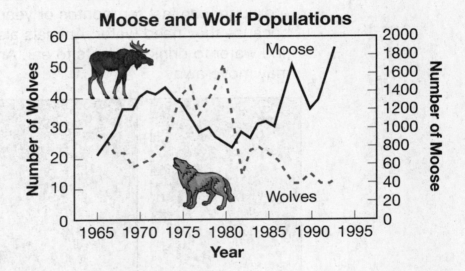

Changes in the Environment

A change in the environment can affect a population. For example, a flood changes the environment quickly. A flood can wash away the homes of plants and animals. It can make the soil too muddy and thin for plants to live in. Then animals that eat those plants cannot find enough food. Some animals may leave the area to look for food and shelter. Others may die.

A fire can change an environment in a few hours. A forest fire kills animals. It also kills trees and other plants. Many animals make their homes in trees. Many animals eat seeds and other parts of plants. When the trees and plants are gone, the animals do not have shelter or food. Animals that survive the fire must leave the area to stay alive.

Some kinds of plants grow quickly after a forest fire. Grass and other small plants cover the burned area quickly. Animals find food and shelter again. But they are not the same plants and animals that lived in the forest before. New populations replaced the old ones.

A dry period called a drought can cause big changes. A drought is a long period of time when very little rain falls. A drought can last for months or years. Many plants die because they need water. Animals also die if they cannot find water to drink or plants to eat. Animals that do not die may move away.

Think about the effects of a wildfire on this forest ecosystem.

When the environment changes, some plants and animals die. Why do some living things survive when others die? Recall that an adaptation is a body part or behavior that helps a living thing stay alive. When the environment changes, some living things have adaptations that help them survive the change. For example, some seeds are able to germinate after they burn. This adaptation helps the plant population survive after a forest fire.

Two sugar maple trees are not exactly alike. Neither are two black bears. Members of a species have differences called **variations**. Variations help some living things survive when the environment changes.

Lesson 43: **Changes in Populations**

DISCUSSION QUESTION

Study the scene below. How would a drought affect these populations?

LESSON REVIEW

1. A fox eats a chipmunk. Which statement is TRUE?

    A. Both the fox and the chipmunk are predators.

    B. Both the fox and the chipmunk are prey.

    C. The fox is the predator and the chipmunk is the prey.

    D. The fox is the prey and the chipmunk is the predator.

2. Suppose that a pond dries up. What will happen to the fish that live in that pond?

    A. They will learn to live on land.

    B. They will not survive.

    C. They will move to another pond.

    D. They will all start to breathe air.

3. Which one is a resource that all living things need?

    A. soil   C. water

    B. seeds   D. trees

Lesson 44: Human Changes in the Environment

LE7.1a, b, c

Getting the Idea

Key Words
- habitat
- deforestation
- pollution
- pollutant
- recycle
- biodegradable

People need resources from their environment. You need food, water, and shelter. People also need space to build homes and natural resources for energy. When people use these resources, people change the environment. Part of our environment is natural. But much of our environment is made by people. Houses and roads are not part of nature. When people make these things, they affect the resources that other living things need.

Habitat Reduction

A **habitat** is the place where a living thing is found naturally. Its habitat is where it lives. An example of a habitat is the part of a forest where a group of deer live. Many different living things share the same habitat.

Humans affect natural habitats in many ways. In order to build new homes and other buildings, people clear land. People change natural habitats into farmland so that they can grow crops and raise animals. The use of land for buildings, mining, or farming is called land development.

When land is developed, people remove many of the nonliving and living resources from the area. Developments often remove the top layer of soil in an area, leaving few or no nutrients for plants. Many habitats around the world have been lost because of development.

The living things in a developed area may be left without food and shelter and often do not survive. The ones that do survive must find new homes, or new areas in which to live. As a habitat gets smaller, it is harder for living things to find the resources they need.

Lesson 44: Human Changes in the Environment

Did You Know?

Scientists believe that some kinds of plastic can last for as long as 50,000 years before they break down. It is good to reuse plastic bags, cups, and containers whenever you can.

One example of natural habitat loss is deforestation. **Deforestation** is the clearing of forest lands. Humans use trees for lumber, food, and products such as paper. After forest land is cleared, trees in the area may grow back. It takes many years for large trees to grow. While the trees slowly grow back, the forest may not support as many living things as it did before.

Pollution

Humans use many natural resources. Natural resources include wood, oil, coal, and gas. Humans change the environment by digging up the land to get oil and coal for energy. The machines and vehicles we use in our businesses, homes, and schools produce chemicals and release them into the water and air. Exhaust made by burning gas and oil in cars, trucks, and planes is also released into the air. Factories often pour their waste into rivers and lakes. We also produce waste that ends up in landfills, causing pollution. **Pollution** is the addition of an unwanted substance into the environment.

A substance that causes pollution is called a **pollutant**. Pollutants often move directly into the air, water, or soil. Many scientists agree that Earth's air is polluted with too much carbon dioxide. Carbon dioxide is a gas that comes from vehicles' exhaust and from electric power plants.

Pollution affects all living things. In some areas, pollution has harmed the soil and water so much that it is hard for plants and animals to live there.

Plastic garbage like these rings from soda cans can be deadly to birds. To keep them safe, garbage must be disposed of properly.

Positive Changes

The good news is that some people are trying to change Earth's environments in good ways. Governments pass laws to help reduce pollution. They also pass laws to protect plants and animals that are hurt by human activities.

Many people recycle parts of their garbage. To **recycle** means to use a material again to make a new product. Recycling helps keep landfills from filling up too quickly. If you recycle at home or at school, you are helping the environment. Your family can also buy things that are made from recycled materials. The sign on the side of the boxes means recycling. This sign on a product lets you know when something contains recycled materials.

Lesson 44: Human Changes in the Environment

Test Tips...

Some questions ask about the definition of a word. Try to define the word before you look at the possible answers. Then look for your definition among the choices.

Using more biodegradable objects is another way that you can help the environment. **Biodegradable** objects can be broken down by nature. These objects can be recycled into nutrients for plants and animals. An old apple core is biodegradable and will get broken down in the ground. A plastic juice box is not biodegradable. It will stay in a landfill for years and years.

It is better to buy a large container of juice instead of many small containers. Then, there is less waste to throw away. Using a paper cup is better than using a plastic cup. But a cup you can wash and use again is the best choice. Choose small things to do each day to help the environment. When small things add up, they help in a big way!

DISCUSSION QUESTION
What are some things people do where you live that affect the environment?

LESSON REVIEW

1. What is the BEST description of a pollutant?
   A. a substance that is biodegradable
   B. a substance that reacts with water
   C. an unwanted substance that is added to the environment
   D. the release of trash into the soil

2. What is deforestation?
   A. removing soil from land
   B. using paper products in a business
   C. clearing forest lands
   D. polluting a forest

3. Which of the following describes land development?
   A. using land for hiking and fishing
   B. using land for state and national parks
   C. preserving land to keep an ecosystem healthy
   D. using land to build new homes and businesses

4. Which of these is a main cause of air pollution?
   A. littering
   B. development
   C. recycling
   D. driving cars and trucks

CHAPTER 10 Review

1 The diagram below shows living things in an environment.

Which living things in the scene are producers?

A   deer and hare
B   trees and grass
C   hare and butterfly
D   butterfly and dragonfly

2 What is the original source of energy for all living things?

A   air
B   water
C   the soil
D   the sun

3 Which of these does a food chain show?

A   how energy passes to different living things
B   how living things hide from predators
C   how plants produce food from sunlight
D   how much food animals eat

4 Which term describes a living thing that feeds mainly on plants?

A producer
B herbivore
C carnivore
D omnivore

5 Which living things break down dead plants and animals?

A herbivores
B scavengers
C predators
D decomposers

6 Which resource do oak trees *not* compete for in their environment?

A water
B food
C light
D space

7 When a predator population's food supply gets larger,

A the predator population gets larger
B the predator population dies
C the predator population gets smaller
D the predator population uses adaptations

8 Which one is *not* a way that human actions can harm the environment?

    A deforestation

    B pollution of air

    C land development

    D using biodegradable objects

9 How do humans depend on their environment? Choose the *best* answer.

    A for food

    B for water

    C for energy

    D for all of the resources above

10 Frogs live in ponds and wetlands, where there is lots of water. What might happen to a frog population if a wetland were drained so new houses could be built?

_____

_____

New York State Coach, Gold Edition, Science, Grade 4

INVESTIGATION 1

How Does Water Change?

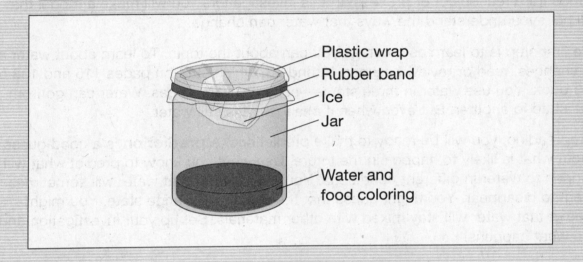

Instructions for This Investigation

Introduction

An **investigation** is a close study of something. Every science investigation starts with a question. Think about water—something you see and use every day. There are many questions you might ask. How does water change form? Where does water go when a puddle dries and disappears? You know that clouds are made of water. So are rain and snow. You know that ice is water, too. But ice has very different qualities compared to the water you drink. Why?

Scientists often use models to investigate things in nature. A **model** is something that stands for a real object or process. In this investigation, you will make a model that will help you understand the ways that water can change.

The first step is to learn as much as you can about the topic. To learn about water and its changes, read or review Lessons 19 and 23, which start on pages 115 and 138 of this book. You use water in three states—solid, liquid, and gas. Water can go from one state to another. But even when it changes, it is still water.

After reading, you will be ready to make predictions. A **prediction** is a good guess about what is likely to happen in the future. Use what you know to predict what will happen to water in different conditions. You may predict that water will sometimes seem to disappear. You might guess that the water will change state. You might predict that water will stay mixed with other materials. Set up your investigation and see what happens!

Materials You Will Need

- large, clear jar
- food coloring
- warm water
- plastic wrap
- rubber band
- ice cubes

Jar

Rubber band

Food coloring

Ice

Steps

1. Fill the bottom of the jar with about 1 cm of very warm water. Add two drops of food coloring.

2. Cut a square piece of plastic wrap. Push it into the top of the jar so that the plastic wrap forms a cup. Make sure all the edges are hanging outside the jar.

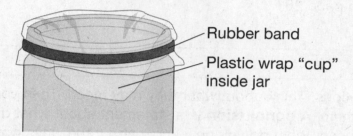

3. Use the rubber band to fasten the plastic wrap tightly in place.

4. You are going to add ice cubes to the plastic wrap cup. But first, make your predictions. What will happen in your setup after you add the ice? You can write your predictions in the chart on page 264.

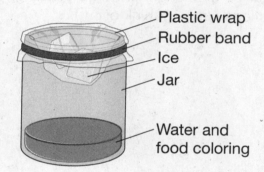

5. Put three ice cubes in the plastic wrap cup in the jar.

6. Observe your model for 10 to 15 minutes. Look closely for any changes that take place. Write down anything you notice.

Observations

As you work, make observations. An **observation** is information that you gather with your senses. Look at the bottom of the plastic wrap cup through the sides of the jar. Also look at the sides of the jar. Notice what happens to the ice. Watch the level of the colored water. Be sure to write down all your observations on a sheet of paper. You can also draw your observations. You will need this information to complete the lab report that starts on page 261.

Conclusion

Study your observations. Think about what they may mean. Then you can draw a conclusion from them. A **conclusion** is a statement about what data from an investigation may mean. What do you now know about the ways water can change? Were your predictions correct? Even if some of your predictions were wrong, you have learned from your investigation.

Scientists test their ideas many times and in different ways. With each test, they learn more. You can do more tests, too. You might test some of water's other qualities. How hot or cold must water be to change form? Do plants give off water? Why does ice float? What other questions can you think of?

New York State Coach, Gold Edition, Science, Grade 4

LAB REPORT FOR INVESTIGATION 1

Name: _____   Date: _____

Questions

What is the topic of your investigation? What are some questions you might ask about it?

_____
_____
_____
_____
_____

Predictions

What do you know about water? Use what you know to make good predictions. What do you think will happen? Write your predictions below.

_____
_____
_____
_____
_____

Materials

What things do you need? List your materials below.

_____
_____
_____
_____
_____

Steps

List the steps of your investigation in order. Be sure to number the steps.

Safety

How did you keep yourself and others safe during your investigation? Write two ways in which you were careful.

Observations

Use your observations to complete the chart below. This chart helps organize your information. If you wish, add a question in the first column.

Predictions and Observations

Question	Prediction	Observations
What will the ice do?		
Will the warm colored water disappear?		
Will drops form inside the jar?		
If drops form, what color will they be?		

Conclusion

What did you learn about water and the ways that it can change? Were your predictions correct? Write your conclusion below.

_____
_____
_____
_____
_____
_____
_____

New York State Coach, Gold Edition, Science, Grade 4

INVESTIGATION 2

Why Does the Moon's Shape Seem to Change?

Instructions for This Investigation

Introduction

If you look at the moon on different nights, you can see that it seems to change shape. These changes in shape are known as the phases of the moon. Why do these changes happen? All good scientists try to make sense of the things they see. Think about what might make the moon seem to change shape.

Scientists often make models to help answer questions. A **model** is something that stands for a real object or system. In this investigation, you will make a model of the Earth-moon-sun system. The model will help you understand and diagram the phases of the moon. Your teacher has set up the classroom as part of the model.

Make a Prediction

The moon goes through phases in which the way the moon looks changes. Why do you think the moon seems to change shape? Make a prediction. A **prediction** is a guess about what is likely to happen or about what is true. Write your prediction in your lab report. Use words and pictures to explain why you think we see phases of the moon.

Do Some Research

Scientists learn as much as they can about the topic they are investigating. After you make your prediction, read or review Chapter 4, "Earth's Movement in Space," which starts on page 89 of this book. Pay extra attention to Lesson 16, "Our View of the Moon" on page 98.

Materials You Will Need

- 8 index cards
- white foam ball
- permanent marker
- pencil

Steps

1. Use the marker to draw a line all the way around the ball, dividing it in half. Color in one-half of the ball. Insert the pencil where the darkened part meets the white. The ball now represents the moon. The darkened part is the area of the moon that is in shadow.

2. Find the model "sun" in your classroom. The white part of your moon model represents the part of the moon that gets light from the sun. You will always keep the "lit" part of the moon model facing toward the model sun.

3. Write a number from 1 to 8 at the top of each index card. You will use these cards for your data. You will face in different directions and record your observations.

4. In this model, you represent Earth. Hold the moon right in front of you, using the pencil as a handle. Always keep the white half of the moon facing the model sun.

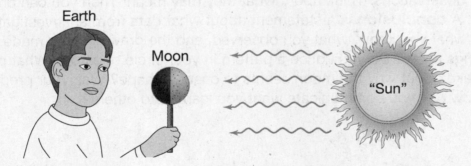

5. Face the location in your classroom labeled with the number 1. Hold the moon in front of you with the light half facing the sun. What does the moon model look like from your point of view? Draw what you see on the index card labeled with the number 1.

6. Turn a little to face the location labeled with the number 2. Turn the moon a little to keep the light half facing the sun. What does the moon model look like now? Draw what you see on the index card labeled with the number 2. Repeat this step for numbers 3 through 8.

Observations

As you work, you will make observations. An **observation** is something that you notice with your senses. Look at how the white and dark parts of the ball seem to change. These changes hold the clues to the phases of the moon. Draw what you observe. After you complete your drawings, arrange the cards on your desk in a circle that matches the numbers located around your classroom. You will need this information to complete the lab report that begins on page 269.

Conclusion

Study your observations. Think about what they may mean. Then you can draw a conclusion. A **conclusion** is a statement about what data from an investigation may mean. Use what you know, what you observed, and the drawings you made. What do your drawings show? Do you notice a pattern in your circle of cards? What conclusion can you make about why the moon seems to change shape? Was your prediction correct? How can you communicate what you learned to others?

New York State Coach, Gold Edition, Science, Grade 4

LAB REPORT FOR INVESTIGATION 2

Name: _____   **Date:** _____

Question

What do you want to find out about the moon in this investigation? Write your question below.

_____

_____

_____

Prediction

Why do you think the moon seems to change shape? Use what you know to make a good guess. Write or draw your prediction below.

_____

_____

_____

Materials

What do you need to do this investigation? List the materials below.

_____

_____

_____

Steps

Write the steps for your investigation. Be sure to number your steps.

_____

_____

_____

_____

_____

_____

_____

_____

_____

_____

Safety

How did you keep yourself and others safe during this investigation? Write about one or two ways.

_____

_____

_____

Observations

Fill in your observations below. Use the index cards you made to fill in the blank spaces in the chart. These cards will help you organize and understand your data.

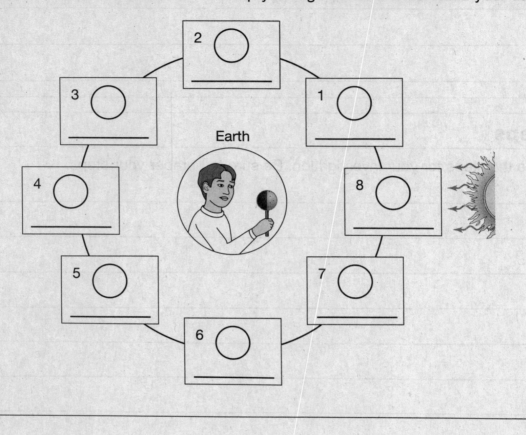

Conclusion

Was your prediction correct? Did you find an answer to your question? What did you learn about the way the moon changes shape? Write your conclusion below.

Glossary

absorb to take in (Lesson 28)

acquired trait a characteristic that a living thing gets during its lifetime. Acquired traits include learned traits. (Lesson 36)

adaptation a body part or behavior that helps a living thing stay alive (Lessons 37, 39)

alive able to move, grow, and respond to the environment (Lesson 35)

analyze to study something carefully in order to understand it (Lessons 8, 13)

anemometer a tool used to measure wind speed (Lesson 18)

area the amount of space an object's surface covers (Lesson 24)

attract to pull closer (Lesson 33)

axis an imaginary line that runs through Earth's center from its North Pole to its South Pole (Lesson 14)

balanced diet a combination of foods that give the body the energy and nutrients that it needs (Lesson 41)

bar graph a chart that shows numbers as bars of different lengths (Lesson 7)

beak the hard, hornlike part of a bird's jaws and mouth (Lesson 39)

behavior a way of acting (Lesson 36)

biodegradable able to be broken down by nature and recycled into nutrients for plants and animals (Lesson 44)

brain the part of an animal's body that learns and thinks (Lesson 39)

bud a plant part that can grow into a new stem, leaf, or flower (Lesson 38)

calendar a chart used to keep track of days, weeks, and months (Lesson 17)

camouflage a coloring or pattern that helps an animal blend in with its surroundings (Lesson 39)

carnivore an animal that eats mainly other animals (Lesson 42)

change of state a change of matter from one form to another, such as liquid to solid (Lesson 23)

chemical change a change in matter in which new materials form (Lesson 27)

chemical energy energy that is stored in a material (Lesson 26)

circuit a path through which electrical energy can flow (Lesson 29)

classify to organize a large group of items into smaller groups, or categories (Lesson 25)

closed circuit a complete path or loop through which electrical energy can flow (Lesson 29)

communicate to share ideas and information (Lesson 9)

community all the populations that live in the same area (Lesson 43)

compound machine a machine made up of two or more simple machines (Lesson 34)

conclusion a statement about what data from an investigation may mean (Lessons 6, 8; Investigations 1, 2)

condensation the process by which a gas changes to a liquid (Lesson 19)

conduction the flow of heat between objects that touch each other (Lesson 27)

conductive able to carry electrical energy (Lesson 22)

conductor a material that heat flows through easily (Lesson 27)

conductor a material that electrical energy can flow through (Lesson 29)

cone a plant part in which pine trees make seeds (Lesson 38)

consumer a person who buys a product (Lesson 12)

consumer an living thing that gets energy by eating other living things (Lesson 42)

controlled variable a condition that is kept the same in all parts of an experiment (Lesson 6)

customary unit a standard unit of measurement used in the United States, such as *inch* or *pound* (Lesson 4)

data pieces of information (Lessons 6, 7)

day the time it takes for Earth to rotate on its axis once (Lesson 17)

decomposer a living thing that gets energy by breaking down dead material from other living things (Lesson 42)

deforestation the cutting down of large areas of forest (Lesson 44)

dependent variable a factor that may change when a scientist changes the independent variable (Lesson 6)

deposition the process by which eroded rock is dropped in new places (Lesson 20)

design to make a plan for solving a problem (Lesson 11)

direction the line or path along which an object moves (Lesson 31)

earthquake the shaking that happens when rock inside Earth breaks and slips (Lesson 21)

electrical energy the energy of moving electric charges (Lessons 26, 29)

energy the ability to do work or cause change (Lesson 26)

engineer a person who works on new technology (Lesson 10)

Glossary

environment a living thing's surroundings (Lesson 43)

equator an imaginary line that divides Earth in half horizontally (Lesson 15)

erosion the process by which weathered rock is picked up and moved to a new place (Lesson 20)

evaluate to judge the value of something (Lesson 13)

evaporation the process by which a liquid changes to a gas (Lesson 19)

exercise physical activity that keeps the body healthy (Lesson 41)

experiment a test performed to answer a scientific question (Lesson 6)

fact a piece of information that is true. A fact is supported by observations. (Lesson 1)

flexible able to bend (Lesson 22)

flood an overflow of water onto land that is usually dry (Lesson 21)

flower a part of a plant in which seeds form (Lesson 37)

food chain a model that shows the path of energy as it flows from one living thing to the next (Lesson 42)

food group a part of a food pyramid (Lesson 41)

force a push or a pull (Lessons 31, 32)

friction a force that acts between two objects that touch or rub together (Lessons 27, 32)

fruit a plant part that forms around seeds (Lesson 37)

fulcrum the support on which a lever rests (Lesson 34)

gas matter that has no definite shape or volume (Lesson 23)

germ a tiny living thing that can cause disease (Lesson 41)

germinate to sprout (Lessons 37, 38)

graduated cylinder a tool for measuring the volume of liquids (Lesson 3)

graph paper a type of paper with lines running across and up and down, used when drawing graphs and tables (Lesson 7)

gravity a force that pulls all objects toward each other. On Earth, gravity pulls objects toward the center of the planet. (Lesson 32)

groundwater water located below Earth's surface (Lesson 19)

habit something you get used to doing all the time. A habit can be good or bad. (Lesson 41)

habitat the place where a living thing lives (Lesson 44)

healthy being strong, having energy, and not getting sick often (Lesson 41)

heat energy flowing from the particles of one object or material to the particles of another object or material (Lessons 26, 27)

hemisphere one half of Earth (Lesson 15)

herbivore an animal that eats mainly plants (Lesson 42)

hibernate to go into a sleeplike state during winter (Lesson 36)

horizon the line where the sky and Earth's surface seem to meet (Lesson 14)

hour a part of a day. A day is divided into 24 hours. (Lesson 17)

humus a material in soil that comes from the remains of dead plants and animals (Lesson 20)

hurricane a huge rainstorm that forms over the ocean and brings very strong winds with it (Lesson 21)

hygiene actions that keep a person clean (Lesson 41)

inclined plane a simple machine that consists of a slanted surface that connects a lower level with a higher level (Lesson 34)

independent variable a condition that a scientist changes in different parts of an experiment (Lesson 6)

inherited trait a characteristic that a living thing gets from its parents before birth (Lesson 36)

instinct an inherited behavior (Lesson 36)

insulator a material through which heat does not flow easily (Lesson 27)

insulator a material through which electrical energy cannot flow (Lesson 29)

interact to act on each other (Lesson 30)

invention something useful that is made for the first time (Lesson 10)

investigation a careful study of something done to answer a question (Lesson 6, Investigation 1)

landform a feature of Earth's surface, such as a mountain (Lesson 20)

larva the second stage in the life cycle of some insects. A caterpillar is the larva of a butterfly. (Lesson 40)

leaf a plant part that captures the sun's energy and makes food for the plant (Lesson 37)

lever a simple machine made up of a bar that rests on a support called a fulcrum (Lesson 34)

life cycle all the stages of a living thing from the beginning of life to death (Lessons 38, 40)

life span the length of time a living thing stays alive (Lessons 38, 40)

light energy that we can see (Lessons 26, 28)

line graph a graph that shows data as a line across a grid (Lesson 7)

liquid matter that has a definite volume but not a definite shape (Lesson 23)

Glossary

magnet an object that can push or pull on some metals without touching them (**Lesson 33**)

magnetic field the area around a magnet where its force can act (**Lesson 33**)

manufacture to make a product (**Lesson 10**)

mass the amount of matter that makes up an object (**Lessons 3, 22, 24, 32**)

matter anything that has mass and takes up space (**Lesson 22**)

measure to find the amount or size of something. When you measure, you compare a known unit to an object or action. (**Lessons 1, 4**)

mechanical energy the energy of moving objects (**Lesson 26**)

metamorphosis the change of form that some animals go through as they become adults (**Lesson 40**)

metric system a system of measurement based on units of ten. Scientists most often use the metric system. (**Lesson 4**)

migrate to move from one place to another in a pattern (**Lesson 36**)

minute a part of an hour. An hour is divided into 60 minutes. (**Lesson 17**)

model something that stands for a real object, process, or system (**Investigations 1, 2**)

month a part of a year. A year is divided into 12 months. (**Lesson 17**)

motion an object's change in position over time (**Lesson 31**)

nutrient a substance that gives a living thing energy and materials for growth (**Lessons 20, 35**)

observation information gathered through the senses (**Lesson 6; Investigations 1, 2**)

observe to gather information using your senses (**Lesson 1**)

offspring the young of a living thing, such as a puppy from adult dogs (**Lessons 35, 36, 40**)

omnivore an animal that eats both plants and other animals (**Lesson 42**)

open circuit a circuit with a broken path, so electrical energy does not flow (**Lesson 29**)

opinion a person's view of or belief about something (**Lesson 1**)

orbit the path of an object in space around another object (**Lesson 15**)

organize to arrange in some kind of order (**Lesson 7**)

pan balance a tool used for measuring mass (**Lesson 3**)

pattern something that repeats, such as a set of events (**Lesson 2**)

phase a shape of the lit half of the moon as seen from Earth (**Lesson 16**)

photosynthesis the process by which plants use the energy of sunlight to make their own food (Lesson 37)

physical change a change in matter in which the kind of matter stays the same. No new materials form. (Lesson 23)

physical property a feature of matter that can be observed with the senses (Lesson 22)

pole one end of a magnet (Lesson 33)

pollen a powdery material that flowering plants use to reproduce (Lesson 38)

pollutant a substance that causes pollution (Lesson 44)

pollution the addition of an unwanted substance into the environment (Lesson 44)

population a group of living things of the same species living in the same place (Lesson 43)

position the place where something is located (Lesson 31)

precipitation water that falls from clouds to Earth's surface, in the form of rain, sleet, hail, or snow (Lessons 18, 19)

predator an animal that hunts other animals for food (Lesson 43)

prediction a guess about what is likely to happen or about what is true (Lessons 2, 6; Investigations 1, 2)

prey an animal that is hunted by other animals for food (Lesson 43)

producer a living thing that makes its own food (Lesson 42)

prototype the first sample of an invention (Lesson 13)

pulley a simple machine made up of a wheel with a rope wrapped around it (Lesson 34)

pupa the stage of an insect's life cycle in which a larva changes into an adult (Lesson 40)

rain gauge a tool used to measure the amount of rain that falls (Lesson 18)

record information written down and saved for future use (Lesson 8)

recycle to use a material again to make a new product (Lesson 44)

reflect to cause to bounce back (Lessons 16, 28)

repel to push away (Lesson 33)

report a detailed description of an investigation (Lesson 9)

reproduce to make more of one's own kind (Lessons 35, 37, 40)

resource something that a living thing needs to stay alive, such as food, water, or air (Lesson 43)

result what is learned from an investigation (Lesson 9)

revision a change in a design (Lesson 12)

Glossary

revolve to move in a path around another object. Earth revolves around the sun. **(Lesson 15)**

root a plant part that takes in water and nutrients from the soil. A plant's roots also hold the plant in the soil. **(Lesson 37)**

rotate to spin on an axis **(Lesson 14)**

runner a stem that grows sideways on the soil surface and forms buds **(Lesson 38)**

runoff water that flows over the surface of the land **(Lesson 19)**

safety goggles a protective covering for the eyes **(Lesson 5)**

safety symbol a small picture that shows possible dangers in a science lab **(Lesson 5)**

screw a simple machine that consists of an inclined plane wrapped around a center pole **(Lesson 34)**

second a part of a minute. A minute is divided into 60 seconds. **(Lesson 17)**

sediment material that is deposited by wind or water **(Lesson 20)**

seed a part of a flowering plant from which a new plant can grow **(Lesson 37)**

sense organ a body part that lets an animal learn about its surroundings **(Lesson 39)**

senses sight, hearing, smell, touch, and taste **(Lesson 1)**

sequence a pattern of events that always happen in the same order **(Lesson 2)**

simple machine a device that makes work easier. A simple machine has few or no moving parts. **(Lesson 34)**

sleep a very important period of rest for the body **(Lesson 41)**

soil the upper layer of Earth's land surface **(Lesson 20)**

solid matter that has a definite shape and a definite volume **(Lesson 23)**

solution an answer to a problem **(Lesson 12)**

sound energy that we can hear **(Lessons 26, 28)**

species a kind of living thing **(Lesson 36)**

speed a measure of how far an object moves in a certain amount of time **(Lesson 31)**

spore a plant cell that can grow into a new plant **(Lesson 38)**

spring scale a tool used to measure weight **(Lesson 3)**

state of matter the form that matter has, such as solid, liquid, or gas **(Lessons 22, 23)**

stem a plant part that moves water and nutrients from the roots to the other parts of the plant. A plant's stem also holds up the leaves, flowers, and fruit. **(Lesson 37)**

sunrise the time of day when the sun appears to come up over the horizon (Lesson 14)

sunset the time of day when the sun appears to fall below the horizon (Lesson 14)

system a group of parts that work together (Lesson 2)

table a chart made up of rows and columns (Lesson 7)

technology the use of science to solve problems and make our lives easier (Lesson 10)

temperature a measure of how warm something is (Lessons 18, 27)

texture the way a surface feels, such as rough or smooth (Lesson 22)

thermometer a tool for measuring temperature (Lessons 3, 18, 27)

thunderstorm a heavy rainstorm that includes lightning and thunder (Lesson 21)

topsoil the upper layer of soil, in which most plants take root and grow (Lesson 20)

tornado a whirling wind that looks like a dark cloud in the shape of a funnel (Lesson 21)

trait a characteristic of a living thing (Lesson 36)

trend a change that happens in a steady, predictable way (Lesson 2)

unit of measurement a standard amount (Lesson 4)

variable something that can change, or vary, in an experiment (Lesson 6)

variations differences among members of a species (Lesson 43)

vibrate to move back and forth very fast (Lesson 28)

volcano an opening in Earth's surface that lets melted rock and gases escape (Lesson 21)

volume the amount of space an object or material takes up (Lessons 3, 22, 24)

waning becoming smaller. The moon is said to be waning when the lit part as seen from Earth is getting smaller. (Lesson 16)

waste a material that a living thing takes in but does not use (Lesson 19)

water cycle the movement of water from Earth's surface into the air and back again (Lesson 19)

waxing becoming larger. The moon is waxing when the lit part as seen from Earth is getting larger. (Lesson 16)

weather the condition of the outside air at a certain time and place (Lesson 18)

weathering the breaking down of rock on Earth's surface into smaller pieces (Lesson 20)

wedge a simple machine that consists of two inclined planes placed back to back (Lesson 34)

week a period of time made up of seven days in a row (Lesson 17)

Glossary

weight a measure of the pull of gravity on an object (Lessons 3, 24)

wheel and axle a simple machine consisting of a wheel fastened to a smaller round bar called the axle. Turning the wheel makes the axle turn. (Lesson 34)

wildfire a fire in a natural area (Lesson 21)

wind vane a tool used to find the direction of the wind (Lesson 18)

work the use of a force to move an object (Lesson 34)

year the time it takes for Earth to complete one orbit around the sun (Lesson 17)

New York State Coach, Gold Edition, Science, Grade 4

PRETEST

Name: _____

Part I

1 One kilogram is equal to

A   1000 grams

B $\frac{1}{100}$ gram

C $\frac{1}{1000}$ gram

D   10 grams

2 As Earth spins, the part facing *away* from the sun has

A   spring
B   night
C   daylight
D   summer

3 Which one is liquid precipitation?

A   rain
B   hail
C   sleet
D   snow

4 The bits of the remains of dead plants and animals in soil are called

A   sediment
B   weathered rock
C   subsoil
D   humus

5 Which object would be attracted to a magnet?

A   rubber eraser
B   wooden spoon
C   iron nail
D   plastic spoon

6 Which one can add sediments and nutrients to soil?

  A hurricane
  B tornado
  C flood
  D earthquake

7 Anything that has mass and takes up space is

  A matter
  B density
  C magnetic
  D conductive

8 What can you use to measure the volume of a liquid?

  A spring scale
  B digital scale
  C balance
  D graduated cylinder

9 Which one is an example of electrical energy?

  A hot stove
  B lightning
  C the sun
  D a drum

10 A push or a pull can change an object's

  A mass
  B temperature
  C magnetic force
  D position

11 What energy change happens when you hit a drum?

    A mechanical energy to sound
    B electrical energy to sound
    C mechanical energy to electrical energy
    D electrical energy to heat

12 You place a book on a shelf. How can you describe the shelf's position?

    A The shelf is behind the book.
    B The shelf is above the book.
    C The shelf is far away from the book.
    D The shelf is underneath the book.

13 The diagram below shows someone shaping a piece of wood. The person is hitting a chisel with a hammer.

What form of energy is the person using?

    A heat energy
    B friction energy
    C mechanical energy
    D sound energy

14 What happens when you touch a hot dish?

    A Heat is transferred from your hand to the dish.

    B Heat is transferred from the dish to your hand.

    C Mechanical energy is transferred from your hand to the dish.

    D Electrical energy is transferred from the dish to your hand.

15 Which one is *not* a way plants reproduce?

    A with seeds
    B with spores
    C with runners
    D with spines

16 Which one is a learned behavior for an animal?

    A hibernating
    B making a nest
    C getting food from trash cans
    D taking care of offspring

17 Which food group contains bread and cereal?

    A grains
    B vegetables
    C fruits
    D milk

18 Which plant part makes most of the food for plants?

    A leaf
    B stem
    C root
    D fruit

19 To learn about danger in its surroundings, an animal uses its

   A jaws
   B sense organs
   C camouflage
   D behaviors

20 Which one correctly shows the life cycle of a frog?

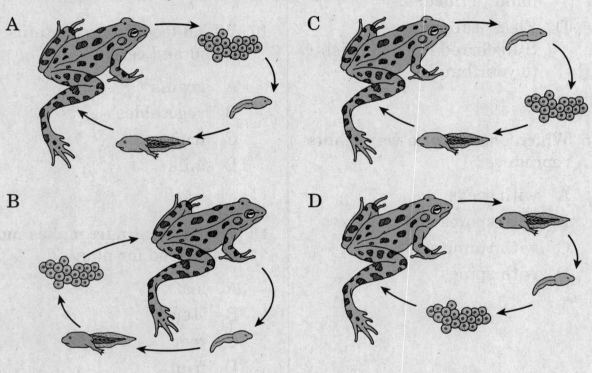

21 Owls eat mice. Which statement describes this relationship?

A Owls are predators, and mice are prey.
B Owls are prey, and mice are predators.
C Both owls and mice are predators.
D Both owls and mice are prey.

22 A student runs down the street on a hot summer day. Which statement is an example of the student's body responding to the environment?

A The student returns home and plays with friends.
B The student plans to run every other day.
C The student enjoys the run.
D The student perspires.

23 Which one is an example of a decomposer?

A mushroom
B sea turtle
C oak tree
D bald eagle

24 Some birds leave northern areas each fall. Then they come back every spring. This adaptation is called

A migration
B hibernation
C defense
D camouflage

25 Which of these is *not* something that all living things do?

A reproduce
B make food
C take in nutrients
D get rid of waste

26 Which statement is an example of how humans have changed the natural environment?

A Wind and water carve a canyon in rock.
B A fire caused by lightning burns a forest.
C A forest is cut down to get wood for building.
D A tornado rips trees out of the ground.

27 What do all plants need from their environment?

A air, food, water, and nutrients
B air, light, water, and humus
C air, light, water, and nutrients
D air, food, water, and fertilizer

28 The length of time that a bird lives after it hatches from an egg is called the bird's

A life cycle
B growth
C life span
D adaptation

29 The drawing below shows a tool used in science.

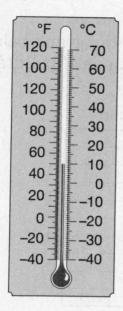

This tool is used to measure

A mass
B density
C volume
D temperature

30 A student observed the birds in his backyard for three days. He wanted to find out which kind of food the birds liked best. Each day he put a different kind of food in his bird feeder. The table below shows the data he collected.

Birds Seen Eating Different Foods

Kind of Food	Number of Birds
peanut butter	6
birdseed	12
suet	18

The student could also use a bar graph to organize his data. How many bars should he use?

A 2
B 3
C 6
D 18

Part II

Base your answers to questions 31 and 32 on the graph below and on your knowledge of science. The graph shows kinds of pollutants that a group of students found on their playground.

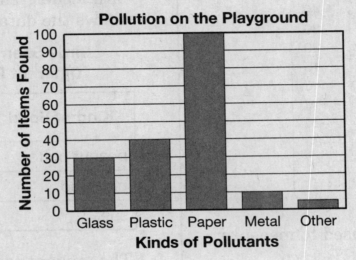

31 List the kinds of pollutants shown in the graph from most items found to fewest items found.

_____

32 The students could also show their data in a table. Use the data in the bar graph to fill in the table below.

Pollution on the Playground

Kind of Pollutant	Number of Items Found
glass	
plastic	
paper	
metal	
other	

Base your answers to questions 33 and 34 on the diagram below and on your knowledge of science. The diagram shows the water cycle.

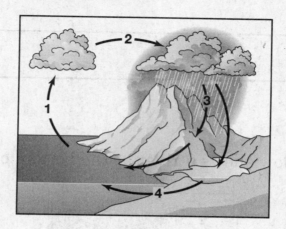

33 Which arrow shows precipitation?

Arrow _____

34 Which arrow shows the part of the water cycle directly caused by heat from the sun?

Arrow _____

35 The diagram below shows an electric circuit.

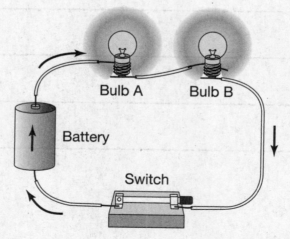

What form of energy inside the battery changes to electrical energy in the wire?

_____ energy

36 Two blocks look just alike. One block sinks, and the other floats. What does that observation tell you about the blocks?

_____

_____

37 The diagram below shows a river at the bottom of a canyon.

How did weathering and erosion work together to form the canyon in the diagram?

_____

_____

_____

_____

38 The diagram below shows six different shapes.

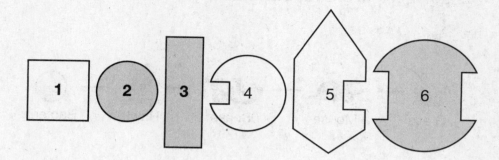

Describe **two** ways you could divide these shapes into two groups.

(1) _____

_____

(2) _____

_____

Base your answers to questions 39 and 40 on the diagram below and your knowledge of science. The diagram represents a food chain.

39 What is the *original* source of all the energy that is transferred through this food chain?

_____

40 Which organism in the food chain is a *producer*?

_____

41 How does a skunk's ability to spray a bad-smelling liquid help it survive?

_____

_____

New York State Coach, Gold Edition, Science, Grade 4

POSTTEST

Name: _____

Part I

1. Which list shows the units in order from smallest to largest?

    A millimeter, kilometer, meter, centimeter

    B meter, kilometer, centimeter, millimeter

    C centimeter, kilometer, meter, millimeter

    D millimeter, centimeter, meter, kilometer

2. Which one is most likely a good conductor of heat?

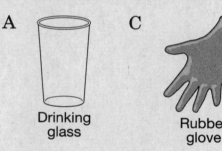

A Drinking glass
B Plastic shovel
C Rubber glove
D Steel can

3. Which one describes a solid?

    A definite shape but no definite volume

    B definite shape and definite volume

    C no definite shape but definite volume

    D no definite shape or definite volume

4. A magnet can attract a steel paper clip

    A through air but not through water or glass

    B through air and water but not through glass

    C through glass but not through air or water

    D through air, water, and glass

5. How long is one year?

    A 7 weeks

    B 24 hours

    C 12 months

    D 24 days

6 An animal's body changes chemical energy in food into

  A electrical energy and sound

  B mechanical energy and light

  C heat and light

  D mechanical energy and heat

7 Which event is caused by gravity?

  A A pencil falls when you drop it.

  B A book slides across a desk when you push it.

  C A ball rolling across a carpet slows down and stops.

  D A bicycle changes direction when you turn the handlebars.

8 Where does the sun set?

  A in the north

  B in the east

  C in the south

  D in the west

Note that question 9 has only three choices.

9 If you move the south pole of one magnet closer to the north pole of another magnet, what happens to the magnetic force between the magnets?

  A The pull between the magnets gets weaker.

  B The pull between the magnets gets stronger.

  C The pull between the magnets stays the same.

10 The diagram below shows a compound machine that can be used to drill holes. It has a screw at one end. A person turns the handle in the middle.

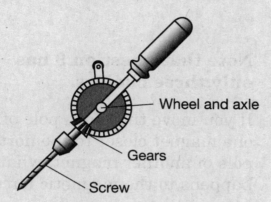

Not all of the mechanical energy used to turn the handle drills the hole. Some of the energy changes to heat. What causes this?

A gravity
B friction
C magnetism
D electricity

11 A student steps outside and notices it is a warm, cloudy day. What is the student observing?

A weather
B climate
C biome
D precipitation

12 What happens if you shine a bright light on your hand?

A Your hand gets cooler.
B Your hand moves.
C Your hand casts a shadow.
D Nothing happens.

13 What makes your hands feel warmer when you rub them together?

A condensation
B friction
C burning
D chemical changes

14 A student plays a violin. What energy transformation takes place?

A light energy to sound energy
B mechanical energy to sound energy
C electrical energy to mechanical energy
D chemical energy to electrical energy

15 Which one needs air, water, and food from its environment?

  A bird
  B tree
  C fern
  D rock

16 An arctic hare has brownish fur in summer. It grows white fur before snow falls in winter. This is an example of

  A learned behavior
  B camouflage
  C hibernation
  D migration

17 Which pair of objects are both *nonliving*?

  A tree and bird
  B tree and birdhouse
  C cloud and bird
  D cloud and plane

18 How do animals get the energy and materials their bodies need?

  A They eat food.
  B They make their own food.
  C They drink water.
  D They get lots of sunlight.

19 Elephants move from place to place to find food and water. This is an example of

  A pollination
  B hibernation
  C migration
  D camouflage

20 Kangaroo rats sleep during the day and are active at night. These animals most likely live in

  A cold, snowy areas
  B hot, dry deserts
  C cool forests
  D cool wetlands

Note that question 21 has only three choices.

21 If two dogs with large ears have puppies, what kind of ears will their offspring most likely have?

   A large ears
   B small ears
   C medium-sized ears

22 A plant's leaves face toward a sunny window nearby. This is evidence that the plant

   A does not like sun
   B needs more water
   C must be kept indoors
   D can respond to its environment

23 Which of these is an example of an inherited trait?

   A a child speaking English
   B a plant having blue flowers
   C a dog having a scar on its ear
   D a raccoon finding food in a trash can

24 Which of these activities will *not* help you stay healthy?

   A washing your hands often
   B getting regular exercise
   C eating a balanced diet
   D smoking cigarettes

25 The length of time from the beginning of a plant's development to its death is called a

   A food chain
   B metamorphosis
   C life span
   D life cycle

26 Which statement is an example of how humans depend on their natural environment?

   A Humans learn to speak different languages.
   B Humans tend to look like their parents.
   C Humans burn fuels to stay warm.
   D Humans use computers to find information.

27 Which physical feature would most likely help an animal survive in a cold, snowy environment?

A short whiskers
B thick, white fur
C dark fur
D a long tail

28 The diagram below shows three dogs. The dogs are all part of the same species, but they look quite different.

Great Dane

Afghan hound

Dachshund

What word describes the dogs' differences?

A adaptations
B senses
C variations
D behaviors

29 Which statement is a prediction that you can test?

A Football is more fun to play than to watch on television.
B The weather in May is not as rainy as the weather in April.
C Vanilla is the best flavor for ice cream.
D The seats on the left side of the classroom are better than the ones on the right side.

30 Which of these is a way to communicate the results of an experiment to others?

A giving an oral report
B taking notes in a science journal
C using your senses to observe
D reading a book on the subject

Part II

Base your answers to questions 31 and 32 on the diagram below and on your knowledge of science. The diagram shows the setup for a student's experiment. One plant receives sunlight, and the other plant does not.

31 Write a scientific question that this experiment could help answer.

_____

_____

32 The student waters the plants each day and observes them closely. Name **two** factors the student could measure.

(1) _____

(2) _____

Base your answers to questions 33 and 34 on the graph below and on your knowledge of science. The graph shows how the water level in a lake changed during the month of June.

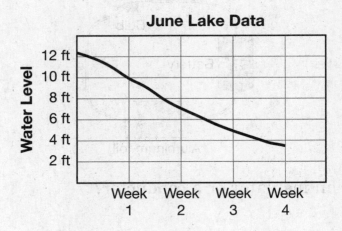

33 Describe how the water level in the lake changed during the month of June.

_____

_____

34 Identify **two** ways that the change might affect living things.

(1) _____

_____

(2) _____

_____

35 The diagram below shows an electrical circuit. The bulb is lit.

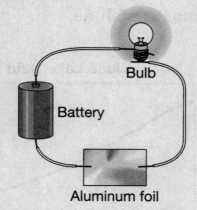

Is aluminum an insulator or a conductor?

_____

36 The diagram below shows ice cubes being heated in a pan. Ice cubes are solid water.

Describe **two** ways the water will change if it is heated enough.

(1) _____

(2) _____

37 Color, texture, odor, hardness, and temperature are all properties of matter. What makes temperature different from the other properties listed?

_____

_____

38 The diagram below shows a soda can in sunlight.

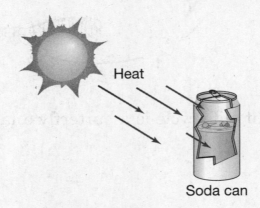

What do you predict will happen to the soda inside the can?

_____

39 The diagram below shows a dandelion plant.

How are a dandelion's seeds most likely spread to new places?

_____

Base your answers to questions 40 and 41 on the diagram below and on your knowledge of science. The diagram shows the life cycle of a butterfly. Each number indicates one stage.

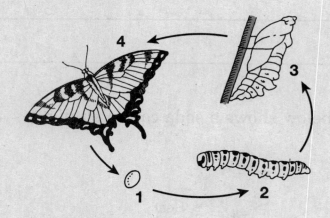

40 At which stage of its life cycle is a butterfly a larva?

Stage _____

41 How does a butterfly change while it is a larva?

_____

Notes

Notes

Notes

Notes